Official Organ of the European Society for Paediatric Endocrinology

Hormone Research

International Journal of Experimental and Clinical Endocrinology

Founded 1970 as 'Hormones' by M. Marois
Continued 1976 by J. Girard (1976–1995)

KARGER

AF882665

Printed in Switzerland on acid-free paper by Reinhardt Druck, Basel

Appears monthly: 2 volumes per year (12 issues)

Guidelines for Authors

Hormone Research

'Hormone Research' aims to provide rapid publication of clinical investigations and case reports, as well as experimental and methodological work related to endocrinology. Correspondence is welcome, but publication of letters is at the discretion of the editors.

Submission

Only original papers written in English are considered and should be sent to:

Prof. M.B. Ranke
S. Karger AG
Editorial Office 'Hormone Research'
P.O. Box
CH–4009 Basel (Switzerland)
Telephone +41 61 306 1344
Fax +41 61 306 1434
E-Mail s.aeschbach@karger.ch

Mailing address for courier deliveries only:

Allschwilerstr. 10
CH–4055 Basel (Switzerland)

Manuscripts should be submitted in triplicate (with three sets of illustrations of which one is an original), typewritten double-spaced on one side of the paper, with a wide margin.

Conditions

All manuscripts are subject to editorial review. Manuscripts are received with the explicit understanding that they are not under simultaneous consideration by any other publication. Submission of an article for publication implies the transfer of the copyright from the author to the publisher upon acceptance. Accepted papers become the permanent property of 'Hormone Research' and may not be reproduced by any means, in whole or in part, without the written consent of the publisher. It is the author's responsibility to obtain permission to reproduce illustrations, tables, etc. from other publications.

Arrangement

Title page: The first page of each paper should indicate the title (main title underlined), the authors' names, and the institute where the work was conducted. A short title for use as running head as well as the full address of the author to whom correspondence should be sent are also required.

Full address: The exact postal address complete with postal code must be given at the bottom of the title page. Please also supply phone and fax numbers, as well as your e-mail address.

Key words: For indexing purposes, a list of 3–10 key words in English is essential.

Abstract: Each paper needs an abstract of up to 10 lines.

Small type: Paragraphs which can or must be set in smaller type (case histories, test methods, etc.) should be indicated with a 'p' (petit) in the margin on the left-hand side.

Footnotes: Avoid footnotes. When essential, they are numbered consecutively and typed at the foot of the appropriate page.

Tables and illustrations: Tables and illustrations (both numbered in Arabic numerals) should be prepared on separate sheets. Tables require a heading and figures a legend, also prepared on a separate sheet. For the reproduction of illustrations, only good drawings and original photographs can be accepted; negatives or photocopies cannot be used. Due to technical reasons, figures with a screen background should not be submitted. When possible, group several illustrations on one block for reproduction (max. size 181 × 223 mm) or provide crop marks. On the back of each illustration, indicate its number, the author's name, and 'top' with a soft pencil.

Color illustrations: Up to 6 color illustrations per page can be integrated with the text at the reduced price of CHF 660.– / USD 545.00 per page. Color illustrations are reproduced at the author's expense.

References: In the text identify references by Arabic numerals [in square brackets]. Material submitted for publication but not yet accepted should be noted as 'unpublished data' and not be included in the reference list. The list of references should include only those publications which are cited in the text. Do not alphabetize; number references in the order in which they are first mentioned in the text. The surnames of the authors followed by initials should be given. There should be no punctuation other than a comma to separate the authors. Cite all authors, 'et al' is not sufficient. Abbreviate journal names according to the Index Medicus system. (Also see International Committee of Medical Journal Editors: Uniform requirements for manuscripts submitted to biomedical journals. N Engl J Med 1997;336:309–315.)

Examples
(a) Papers published in periodicals: Sun J, Koto H, Chung KF: Interaction of ozone and allergen challenges on bronchial responsiveness and inflammation in sensitised guinea pigs. Int Arch Allergy Immunol 1997;112:191–195.
(b) Monographs: Matthews DE, Farewell, VT: Using and Understanding Medical Statistics, ed 3, revised. Basel, Karger, 1996.
(c) Edited books: Parren PWHI, Burton DR: Antibodies against HIV-1 from phage display libraries: Mapping of an immune response and progress towards antiviral immunotherapy; in Capra JD (ed): Antibody Engineering. Chem Immunol. Basel, Karger, 1997, vol 65, pp 18–56.

Page charges

There is no page charge for papers of 4 or fewer printed pages (including tables, illustrations and references). Each additional complete or partial page is charged to the author at CHF 650.– / USD 540.00. The allotted size of a paper is equal to approx. 12 manuscript pages (including tables, illustrations and references). As the page charges are waived in any case for the first paper each year to be published by a member of ESPE, authors who are members of the ESPE should so indicate on the title page.

Galley proofs

Unless indicated otherwise, galley proofs are sent to the first-named author and should be returned with the least possible delay. Alterations made in galley proofs, other than the correction of printer's errors, are charged to the author. No page proofs are supplied.

Reprints

Order forms and a price list are sent with the galley proofs. Orders submitted after the issue is printed are subject to considerably higher prices.

KARGER

Fax+ 41 61 306 12 34
E-Mail karger@karger.ch
www.karger.com

The Guidelines for Authors are available at:
www.karger.com/journals/hre/hre_gl.htm

General Information

Hormone Research

Publication data: 'Hormone Research' is published 12 times a year. Volumes 53 and 54 with 6 issues each appear in 2000.

Subscription rates: Subscriptions run for a full calendar year. Prices are given per volume.
Personal subscription:
Subscription + postage and handling
CHF 308.– + CHF 25.90 Europe
CHF 308.– + CHF 39.10 Overseas
USD 235.50 + USD 25.90
+ online subscription per volume:
CHF 39.– / USD 30.00
(Must be in the name of, billed to, and paid by an individual. Order must be marked 'personal subscription'.)
Institutional subscription:
Subscription + postage and handling
CHF 616.– + CHF 32.30 Europe
CHF 616.– + CHF 48.80 Overseas
USD 471.00 + USD 32.30
+ online subscription per volume:
CHF 58.– / USD 45.00
Airmail postage: CHF 36.–, USD 30.00.
(Extra per volume.)
Discount subscription prices:
European Society for Paediatric Endocrinology. Lawson Wilkins Society; ISGD; Schweizerische Gesellschaft für Endokrinologie.

Change of address: Both old and new address should be sent to the subscription source.

Microforms: Please direct all enquiries and orders to: University Microfilms Inc.
300 North Zeeb Road, P.O. Box 1346
Ann Arbor, MI 48106-1346 (USA)

Subscription orders: Orders can be placed at agencies, bookstores, directly with the Publisher
S. Karger AG, P.O. Box,
CH–4009 Basel (Switzerland)
or with any of the following distributors:
Deutschland: S. Karger GmbH, Postfach
D–79095 Freiburg
France: Librairie Luginbühl
36, bd de Latour-Maubourg, F–75007 Paris
Great Britain, Ireland: S. Karger Publishers
58 Grove Hill Road
Tunbridge Wells, Kent TN1 1SP
USA: S. Karger Publishers, Inc.
26 West Avon Road, P.O. Box 529
Farmington, CT 06085
India, Bangladesh, Sri Lanka:
S. Karger India Subscription Agency
B–5/132 Safdarjung Enclave
New Delhi 110029
Japan: Katakura Libri, Inc.
2F, Marushima Bldg, 4-6-17, Yushima,
Bunkyo-ku, Tokyo 113-0034

Single issues and back volumes: Information can be obtained through the Publisher.

Advertising: Correspondence and rate requests should be addressed to the Publisher.

Bibliographic indices: This journal is regularly listed in bibliographic services, including *Current Contents®*.

Book Reviews: Books are accepted for review by special agreement.

Copying: This journal has been registered with the Copyright Clearance Center (CCC), as indicated by the code appearing on the first page of each article.
For readers in the US, this code signals consent for copying of articles for personal or internal use, or for the personal or internal use of specific clients, provided that the stated fee is paid per copy directly to
Copyright Clearance Center Inc.
222 Rosewood Drive
Danvers, MA 01923 (USA)
A copy of the first page of the article must accompany payment. Consent does not extend to copying for general distribution, for promotion, for creating new works, or for resale. In these cases, specific written permission must be obtained from the copyright owner,
S. Karger AG,
P.O. Box, CH–4009 Basel (Switzerland).

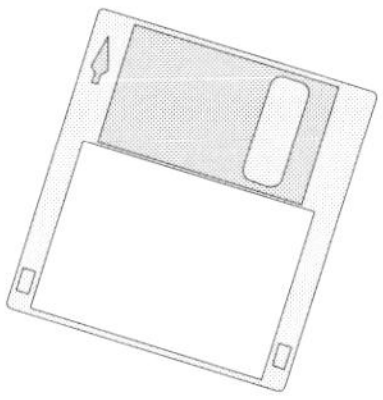

Disk Submission

A **comprehensive version** of these guidelines for the **submission of manuscripts and illustrations on disk** can be accessed online under **www.karger.com/journals/disksub.htm** or requested as a leaflet from the editorial department.

Please follow the general rules on style and arrangement in the **'Guidelines for Authors'** of the journal. Submit a **double-spaced printout** of your manuscript matching the file on disk exactly.

Technical Information
Store only data pertinent to this manuscript. Include your fonts (screen and printer fonts for Macintosh).
For a **text only** manuscript, use a 3.5-inch disk.
For **text and illustrations,** use 3.5 or 5.25 Magneto Optical Disc; CD-ROM (DOS or hybrid); iomage Zip/DOS (100 MB), or 44 MB, 88 MB or 200 MB SyQuest. Compress extensive data using ZIP format.

Data Description
Label your disk with the **title of the journal or book** for which it is submitted; a **short version** of the paper's title; your **name,** postal and e-mail address, fax and phone number; **hard and software** used, including the version; and the **complete list of the file names.** File names consist of eight characters before the dot identifying the file and three characters after the dot identifying the application.

Text and Format
The **preferred word-processing package is MS-Word.** Other commonly used PC text programs and ASCII formats are accepted.
Update your data when you revise subsequent to the reviewer's or editor's decision and mark revisions clearly. Manuscripts accepted for publication are subject to **copy editing:** changes and/or questions will be marked in the manuscript, which will be returned to you with your disk and the galley proofs for checking.
Enter your text continuously flush left. Do not use any layout functions. Do not split words at the end of a line. **Do not use indentations.** Use a blank line before and a hard return at the end of paragraphs.
Arrange headings flush left. Insert only **one space after punctuation marks.** Insert Greek letters, mathematical symbols etc. with your word-processing program. Use **boldface and italics** as well as sub- and superscript where appropriate.
Tables are part of the text and should be placed at the end of the text file. Separate the individual parts of a table by lines. Use the tabulator instead of spaces. Format operation data on one line in the same column. **Legends to illustrations are part of the text** and should also be listed at the end of the text file.

Illustrations
Store illustrations as separate files: do not integrate them in your text. Export **line drawings/vector graphics** as TIF, EPS or WMF format. Use Photoshop for processing and retouching **scanned halftone images.** Save the original scan and the processed version. Export **black and white or color images** as TIF or EPS format in their anticipated size in print.
For detailed information on how to submit illustration data on disk please consult the leaflet or **www.karger.com/journals/disksub.htm.**

KARGER

Fax +41 61 306 12 34
E-Mail karger@karger.ch
www.karger.com

The Journal Home Page is available at:
www.karger.com/journals/hre

KI 98 DISK

A practical handbook for use at the patient's bedside

Practical Algorithms in Pediatric Endocrinology

Editor: Ze'ev Hochberg

Main Headings

- Growth
- Puberty
- Intersex
- Adrenal
- Water and Electrolytes
- Calcium Metabolism
- Thyroid
- Carbohydrates

Medical textbooks are mainly oriented by body systems, disease or diagnosis, yet the practicing physician is confronted with the patient's complaint by a symptom, physical sign or laboratory abnormality, from which he is expected to diagnose and proceed with treatment. The traditional medical approach is through differential diagnosis by exclusion. Algorithms provide a direct approach to breaking down long lists/tables of differential diagnosis into smaller, more manageable ones. Often, a whole group of diagnoses can be excluded by single or a group of signs, blood tests or imaging.

This book is meant as a pragmatic text to be used at the patient's bedside. It classifies common clinical symptoms, signs and laboratory abnormalities as they present themselves in daily practice. Aimed at an audience of general and family practitioners, trainees or pediatricians who are not exposed on a day-to-day basis to pediatric endocrine problems, it provides a logical, concise and cost-effective approach from which they can profit and acquire medical reasoning.

Practical Algorithms in Pediatric Endocrinology
Editor: Hochberg, Z. (Haifa)
IV + 110 p., 52 graphs, 4 fig., 1 tab., spiral bound, 1999
CHF 55.– / DEM 66.– / USD 48.00
Prices subject to change
DEM price for Germany, USD price for USA only
ISBN 3-8055-6693-X

Fields of Interest: Pediatrics; Endocrinology

KI 98680

KARGER

Hormone Research

3rd Serono Symposia Workshop on Clinical Paediatric Endocrinology

Verona, Italy, September 23–24, 1999

SERONO SYMPOSIA

Guest Editors
K. Albertsson-Wikland, Göteborg
J. Argente, Madrid
J.L. Chaussain, Paris
H. Frisch, Vienna
U.E. Heinrich, Heidelberg
M.O. Savage, London
L. Tatò, Verona

35 figures and 15 tables, 2000

KARGER

Basel · Freiburg · Paris · London · New York ·
New Delhi · Bangkok · Singapore · Tokyo · Sydney

This publication was supported by an educational grant from Serono Symposia

S. Karger
Medical and Scientific Publishers
Basel · Freiburg · Paris · London
New York · New Delhi · Bangkok
Singapore · Tokyo · Sydney

Drug Dosage
The authors and the publisher have exerted every effort to ensure that drug selection and dosage set forth in this text are in accord with current recommendations and practice at the time of publication. However, in view of ongoing research, changes in government regulations, and the constant flow of information relating to drug therapy and drug reactions, the reader is urged to check the package insert for each drug for any change in indications and dosage and for added warnings and precautions. This is particularly important when the recommended agent is a new and/or infrequently employed drug.

Printed in Switzerland on acid-free paper by
Reinhardt Druck, Basel
ISBN 3–8055–7097–X

KARGER

Fax+ 41 61 306 12 34
E-Mail karger@karger.ch
www.karger.com

Vol. 53, Suppl. 1, 2000

Hormone Research

Contents

KARGER

Fax+ 41 61 306 12 34
E-Mail karger@karger.ch
www.karger.com

Access to full text and tables of contents, including tentative ones for forthcoming issues:
www.karger.com/journals/hre/hre_bk.htm

Hormone Research

Preface

I am pleased to have the honour of introducing the proceedings of the 3rd Workshop on Clinical Paediatric Endocrinology, which took place in my home town – the beautiful city of Verona – in September 1999. The workshop was attended by many eminent specialists in the field of clinical paediatric endocrinology, whose valuable contributions generated a high level of discussion. I am positive that all those who attended the workshop found it beneficial and enjoyable. These proceedings capture the essence of the workshop, and draw together the current thinking in clinical paediatric endocrinology. These proceedings will be a valuable resource of information in the area.

Luciano Tatò
Workshop Chairman
University of Verona
Italy

KARGER
Fax +41 61 306 12 34
E-Mail karger@karger.ch
www.karger.com

Accessible online at:
www.karger.com/journals/hre

Hormone Research

Horm Res 2000;53(suppl 1):2–6

Loss of Imprinted Genes and Paternal SUR1 Mutations Lead to Focal Form of Congenital Hyperinsulinism

Jean-Christophe Fournet[a,b] Christine Mayaud[a] Pascale de Lonlay[a,c] Virginie Verkarre[a,b] Jacques Rahier[f] Francis Brunelle[d] Jean-Jacques Robert[c] Claire Nihoul-Fékété[e] Jean-Marie Saudubray[c] Claudine Junien[a]

[a]INSERM UR 383 Hôpital Necker-Enfants Malades, Clinique Maurice-Lamy, Paris, and Departments of [b]Pathology, [c]Pediatrics, [d]Radiology and [e]Surgery, Hôpital Necker-Enfants Malades, Paris, France; [f]Department of Pathology, University Hospital St-Luc of Louvain, Brussels, Belgium

Key Words

Alleles, loss · Beckwith-Wiedemann syndrome · Neonatal hypoglycaemia · Persistent hyperinsulinism · Potassium-ATP channel

Abstract

Persistent hyperinsulinaemic hypoglycaemia of infancy (PHHI) is a heterogeneous disorder characterized by profound hypoglycaemia due to inappropriate hypersecretion of insulin. An important diagnostic goal is to distinguish patients with a focal hyperplasia of islet cells of the pancreas (FoPHHI) from those with a diffuse abnormality of islets (DiPHHI), because the management differs significantly. The intriguing similarity between islet cell hyperplasia and tumourigenesis prompted us to investigate whether the imprinted genes in the 11p15 region are involved. Results showed that diffuse forms are caused by constitutional homozygous or compound heterozygous mutations of the SUR1 gene. In contrast, focal forms are caused by loss of the maternally inherited 11p15 region, resulting in both loss of the maternally expressed tumour suppressor genes accounting for hyperplasia and somatic reduction to hemizygosity or homozygosity of the paternally inherited SUR1, limited to the lesion. Thus, this somatic disorder, which leads both to β-cell proliferation and to hyperinsulinism, can be considered the somatic equivalent, restricted to a microscopic focal lesion, of constitutional uniparental disomy associated with unmasking of a heterozygous parental mutation.

Introduction

Persistent hyperinsulinaemic hypoglycaemia of infancy (PHHI; OMIM 256450) is a glucose metabolism disorder characterized by unregulated secretion of insulin and profound hypoglycaemia. The incidence of PHHI is 1 in 50,000 live-births [1]; 95% of the cases are sporadic. Treatment of patients with diazoxide and/or somatostatin analogues is not always effective, necessitating an intervention such as pancreatectomy [2]. Rare familial forms

KARGER

Fax + 41 61 306 12 34
E-Mail karger@karger.ch
www.karger.com

Accessible online at:
www.karger.com/journals/hre

Prof. C. Junien
INSERM UR 383, Hôpital Necker-Enfants Malades
Clinique Maurice Lamy, 149, rue de Sèvres
F–75743 Paris, Cedex 15 (France)
Tel. +33 1 44 49 44 85, Fax +33 1 47 83 32 06, E-Mail junien@ceylan.necker.fr

may be caused by recessive or dominant defects in four different genes: (1) the sulphonylurea receptor gene (SUR1) [3–5]; (2) the inward rectifying potassium channel subunit gene (KIR6.2) [6, 7]; (3) the glutamate dehydrogenase gene (GLUD-1) [8], and (4) the glucokinase gene (GK) [9].

Two types of histopathological lesion are associated with PHHI despite there being one standard clinical presentation: a focal form (FoPHHI) and a diffuse form (DiPHHI) [10–12]. FoPHHI represents approximately 30% of operated cases [12].

Hyperinsulinaemia with hyperplasia of islets of Langerhans may be associated with several overlapping syndromes predisposing to tumours including the Beckwith-Wiedemann syndrome (BWS) (MIM.130650) [13]. The 11p15.5 chromosome region involved in BWS contains an imprinted domain, including several imprinted genes characterized by mono-allelic expression [14–17]. These include: (1) four maternally expressed genes – H19 (a candidate tumour suppressor gene) [17, 18], P57KIP2 (a negative regulator of cell proliferation) [19], KVLQT1 (the gene coding for the potassium channel involved in the long QT syndrome) [15], and HASH2 (a transcription factor) [20] – and (2) one paternally expressed gene – the insulin-like growth factor 2 [14, 21].

However, some genes in this domain, including KVLQT1 and the insulin gene, INS, may escape functional imprinting in some tissues or developmental stages [22]. Alteration of imprinting in BWS in most embryonal tumours and even in adult tumours can result in unbalanced expression affecting either one gene or the whole imprinted domain by loss of the maternal chromosome or relaxation of genomic imprinting [13, 23].

The intriguing similarity between islet cell hyperplasia and tumourigenesis prompted us to investigate whether the genes in the 11p15 region are involved in sporadic cases of PHHI. As it is possible that hyperplasia of islet-like cell clusters is clonal in origin, we searched for loss of alleles in the 11p15 region in 15 cases of FoPHHI and in 6 cases of DiPHHI.

Patients and Methods

Patients with neonatal PHHI onset (within 72 h after birth) and who were diazoxide resistant were investigated. Patients were screened by selective pancreatic venous sampling using transparietal portal catheterization [24] and by peroperative surgical examination and analysis of extemporaneous frozen sections in order to identify patients who had a focal lesion of the islet cells and in whom partial pancreatectomy could be performed [12]. In patients showing a focal lesion, a complete relief of hyperinsulinaemic hypoglycaemia was obtained, and the postoperative plasma glucose levels were strictly normal after a mean follow-up period of 33.4 (range 21–45) months in patients in whom partial pancreatectomy was performed.

Histological diagnosis of focal adenomatous hyperplasia was performed on extemporaneous frozen sections as described and further confirmed after fixation and paraffin embedding [12].

Polymerase chain reaction (PCR) analysis was performed for LOH: 15 microsatellite markers for chromosome 11 were used to genotype paired samples of DNA extracted from frozen samples of normal pancreas and hyperplastic islet cells and of leukocytes of patients and their parents. PCR primer pairs used to amplify the regions coding for the nucleotide binding fold 2 (NBF2) and the NBF1 domain of the SUR1 gene were as described previously. PCR and single-strand conformation polymorphism analyses were performed, and samples containing aberrantly migrating fragments were directly sequenced.

Results

Patients

Twenty-one infants with neonatal diazoxide-resistant PHHI underwent pancreatectomy. Fifteen infants were FoPHHI and 6 were DiPHHI, as assessed by selective pancreatic venous sampling and peroperative surgical examination and analysis of extemporaneous frozen sections.

Pancreatic Histopathology

Partial or near-total pancreatectomy specimens from all 21 patients were analyzed. The two forms gave clearly different patterns. The adenomatous hyperplasia (FoPHHI) was characterized by focal hyperplasia of some islet-like cells, including hypertrophied insulin-secreting cells with giant nuclei. All islets outside the lesion had an apparently normal aspect. In contrast, all the islets of Langerhans throughout the pancreas in the DiPHHI cases were irregular in size and contained distinctly hypertrophied insulin cells. DNA was extracted from acutely sliced frozen samples of hyperplastic and apparently normal pancreas. The focal lesion represented between 50 and 80% of the cells.

Loss of Maternal Alleles of Imprinted Genes in FoPHHI

To examine whether hyperinsulinism was associated with an LOH in the 11p15 region, patients were genotyped for 15 markers along the short arm of chromosome 11. Paired samples of DNA extracted from frozen sections of normal pancreas and hyperplastic islet cells and from leukocytes, were investigated. LOH was observed in islet cell hyperplasia samples from all 15 patients with

FoPHHI. The same polymorphic loci on 11p were examined for LOH in paired constitutional DNA and abnormal pancreas DNA from the 6 DiPHHI patients. No LOH was detected. Thus, unlike FoPHHI, constitutional heterozygosity was consistently maintained in cases of DiPHHI. LOH was often observed as partial losses of alleles. (This may have been due to contamination of islet cell hyperplasia samples with exocrine pancreas cells, some of which, in spite of careful selection on frozen sections, contaminated the samples used for DNA preparation.)

To identify the parental origin of the chromosome lost in FoPHHI cases, the parents were genotyped using polymorphic microsatellite markers covering the whole length of chromosome 11. In all 15 FoPHHI cases, the allele lost in hyperplastic islet cells was of maternal origin. To determine the extent of the LOH, we genotyped the DNA from paired samples of normal pancreas or leukocytes and of hyperplastic tissue using 21 microsatellite markers covering the whole length of chromosome 11. Iso-allelism extended over the entire chromosome in 3 cases (Fo3, Fo4 and Fo5). In 4 cases (Fo1, Fo2, Fo7 and Fo8), LOH did not extend to the telomere of the long arm. The smallest region of overlap of LOH was restricted to the region from the most telomeric marker, D11S922, to D11S899 in case Fo8. This somatic event is consistent with a proliferative monoclonal lesion. To test whether the LOH only affected chromosome 11, we also genotyped paired DNA samples of normal and hyperplastic tissue from all FoPHHI patients using eight polymorphic markers for different loci or regions, some of which have been implicated in insulinoma. Germline heterozygosity was retained for all these markers. The only exception was the MEN1 locus on 11q13, consistent with the complete loss of maternal chromosome 11 in 5 out of 15 patients.

Detection of Heterozygous Mutations in the SUR1 Paternal Allele

The most commonly deleted region of chromosome 11 contained the SUR1 [25] and KIR6.2 [26] genes that code for two subunits of the potassium-ATP channel and are involved in familial recessive PHHI [3–7]. To understand whether the LOH may also unmask a recessive phenotype responsible for hypersecretion of insulin, we looked for deleterious mutations in the SUR1 gene. As most published mutations were located in the NBF2 domain, we studied the exons coding for this domain and the exons coding for the NBF1 domain of the SUR1 gene. Using single-strand conformation polymorphism analysis, we detected a mobility shift by electrophoresis in constitutional DNA from leucocyte samples and from normal pancreas samples from 4 out of 15 patients, Fo1, Fo9, Fo11 and Fo12. In all 4 cases, the mobility shift was also found in leucocyte DNA from their respective fathers. In agreement with the loss of maternal alleles, the signal of the mutant band inherited from the father was enhanced in the DNA from the hyperplastic lesion, while the signal of the normal wild-type band inherited from the mother was greatly diminished. The same pattern was observed for all 13 patients.

Characterization of the SUR1 Gene Mutations

Mutations in the SUR1 (4 patients) genes displaying the band shifts discussed above were identified by nucleotide sequence analysis of PCR products amplified in independent reactions from leucocyte DNA samples. All 4 patients were constitutionally heterozygous for mutations.

In all cases, the father was also constitutionally heterozygous for the same mutation found in the proband. A total of 212 chromosomes from a population of 106 unrelated individuals were tested for the presence of these newly described mutations. In this population and in other individuals reported in the literature, no such missense mutations were detected, precluding a polymorphism. Furthermore, in all 4 cases arginine, a basic amino acid, was replaced by a non-polar amino acid such as proline (position 1352), cysteine (position 1420), or tryptophan (position 1493), increasing the likelihood that the mutation would result in a change in the protein conformation. Finally, comparison of homologous sequences of five different members of the ATP binding cassette protein superfamily revealed that positions 1352, 1420 and 1493 are well conserved and code for an arginine or another basic amino acid. Taken together, these data strongly suggest that these missense mutations are deleterious. In total we found four SUR1 mutations of NBF domains in 15 focal cases.

Discussion

We describe for the first time the occurrence of a recessive endocrine disorder which is due to somatic reduction to hemizygosity or homozygosity of a paternally inherited mutation limited to a focal hyperplastic lesion in 4 patients. The minimal region of LOH includes candidate genes such as the SUR1 gene and the KIR6.2 gene in 11p15.1 and the imprinted domain H19-IGF2-HASH2-KVLQT1-P57KIP2 in 11p15.5.

In the focal lesion, a loss of maternal alleles, therefore, resulted simultaneously in two different consequences – hyperinsulinism and proliferation:

(1) The reduction to hemi- or homozygosity of the paternal defective allele of the SUR1 gene with loss of the wild-type maternal allele responsible for hyperinsulinism. Hyperinsulinism is the direct consequence of loss of function of the potassium-ATP channel as already shown by patch-clamp studies [27]. In all cases the father carried the mutant allele, therefore, excluding the hypothesis of preferential paternal de novo germline mutation.

(2) The loss of maternal alleles includes the cluster of imprinted genes in the 11p15.5 region [28] comprising growth factors such as insulin-like growth factor 2 [14] and tumour suppressor genes such as H19 [18] and P57KIP2 [16]. This necessarily leads to altered expression of these and other genes in the imprinted domain, suggesting that unbalanced expression of the corresponding gene products may give rise to the increase in proliferation of β-cells, a striking feature of focal adenomatous hyperplasia not observed in the diffuse form [29]. Such altered expression of imprinted genes is already well documented for embryonal tumours and remarkably in Wilms' tumour. This is consistent with the lesion being a tumourigenic process as suggested by its morphological features.

Thus, maternal LOH with somatic reduction to homozygosity of the paternal SUR1 mutation can be considered as the somatic equivalent, restricted to this microscopic focal lesion, of constitutional uniparental disomy associated with unmasking of a heterozygous parental mutation leading to a recessive disorder [30]. The genetic alteration causing FoPHHI may thus be similar to that causing BWS, but is limited to the clonal proliferation originating from one islet cell precursor. In BWS, however, hyperinsulinaemia is transient, but it is unclear whether the lesions are focal or diffuse. As demonstrated by antibodies against proliferating cell nuclear antigen, a nuclear antigen marker for cells in S phase, proliferating cells are present in focal lesions but absent in diffuse lesions [31]. There are several examples of abnormal phenotypes associated with uniparental disomy that can result both from the presence of imprinted genes on the chromosome involved in this non-mendelian inheritance mechanism and/or from reduction to homozygosity of a mutation associated with an autosomal recessive disorder [30].

The mechanism involved in FoPHHI is similar to that recently reported for type I autosomal dominant polycystic kidney disease [32]. In contrast to this recurrent loss of 16p in polycystic kidney disease, the somatic loss of 11p15 probably occurs very rarely in heterozygous individuals as demonstrated by the rarity of relapse after partial pancreas resection and by the fact that in individuals heterozygous for SUR1 mutations, hyperinsulinism occurs sporadically with no other case in the family, even though half of the children, the father and possibly other relatives must be carriers of the same mutation in the SUR1 gene. However, in order to define more precisely the risk of occurrence in these populations, but also the risk of recurrence in families with a sporadic case, a systematic search for the mutation in the proband and in relatives would have to be undertaken, together with a retro- or prospective search for focal lesions.

Despite having similar clinical presentation, an important diagnostic goal is to distinguish patients with focal adenomatous hyperplasia of islet cells from those with the diffuse abnormality, because management strategies differ significantly: medical treatment of patients with diazoxide and/or somatostatin analogues is frequently ineffective, necessitating a 95% pancreatectomy in diffuse forms but only partial pancreatectomy in focal forms [8, 12, 33]. According to a mean follow-up period of 6 years in 22 patients with FoPHHI, the somatic nature of the second hit explains that it is possible to perform only partial pancreatectomy limited to the lesion, avoiding the occurrence of iatrogenic diabetes.

References

1 Bruining G: Recent advances in hyperinsulinism and the pathogenesis of diabetes mellitus. Curr Opin Pediatr 1990;2:758–765.
2 Shilyansky J, Fisher S, Cutz E, Perlman K, Filler R: Is 95% pancreatectomy the procedure of choice for treatment of persistent hyperinsulinemic hypoglycemia of the neonate? J Pediatr Surg 1997;32:342–346.
3 Nestorowicz A, Wilson BA, Schoor KP, Inoue H, Glaser B, Landau H, Stanley CA, Thornton PS, Clement JP 4th, Bryan J, Aguilar-Bryan L, Permutt MA: Mutations in the sulfonylurea receptor gene are associated with familial hyperinsulism in Ashkenazi Jews. Hum Mol Genet 1996;5:1813–1822.
4 Thomas PM, Cote GJ, Wohllk N, Haddad B, Mathew PM, Rabl W, Aguilar Bryan L, Gagel RF, Bryan J: Mutations in the sulfonylurea receptor gene in familial persistent hyperinsulinemic hypoglycemia of infancy. Science 1995;268:426–429.
5 Thomas PM, Wohllk N, Huang E, Kuhnle U, Rabl W, Gagel RF, Cote GJ: Inactivation of the first nucleotide-binding fold of the sulfonylurea receptor, and familial persistent hyperinsulinemic hypoglycemia of infancy. Am J Hum Genet 1996;59:510–518.
6 Thomas P, Ye Y, Lightner E: Mutation of the pancreatic islet inward rectifier Kir6.2 also leads to familial persistent hyperinsulinemic hypoglycemia of infancy. Hum Mol Genet 1996;5:1809–1812.
7 Nestorowicz A, Inagaki N, Gonoi T, Schoor K, Wilson B, Glaser B, Landau H, Stanley C, Thornton P, Seino S, Permutt M: A nonsense mutation in the inward rectifier potassium channel gene, Kir6.2, is associated with familial hyperinsulinism. Diabetes 1997;46:1743–1748.
8 Stanley CA, Lieu YK, Hsu BY, Burlina AB, Greenberg CR, Hopwood NJ, Perlman K, Rich BH, Zammarchi E, Poncz M: Hyperinsulinism and hyperammonemia in infants with regulatory mutations of the glutamate dehydrogenase gene. N Engl J Med 1998;338:1352–1357.
9 Glaser B, Kesavan P, Heyman M, Davis E, Cuesta A, Buchs A, Stanley C, Thornton P, Permutt A, Matschinsky F, Herold K: Familial hyperinsulinism caused by an activating glucokinase mutation. N Engl J Med 1998;338:226–230.
10 Goossens A, Gepts W, Saudubray J, Bonnefont J, Nihoul-Fékété C, Heitz P, Kloppel G: Diffuse and focal nesidioblastosis: A clinicopathological study of 24 patients with persistent neonatal hyperinsulinemic hypoglycemia. Am J Surg Pathol 1989;13:766–775.
11 Jaffé R, Hashida Y, Yunis E: Pancreatic pathology in hyperinsulinemic hypoglycemia of infancy. Lab Invest 1980;42:356–365.
12 Rahier J, Sempoux C, Fournet J-C, Poggi F, Brunelle F, Nihoul-Fékété C, Saudubray J-M, Jaubert F: Partial or near-total pancreatectomy for persistent neonatal hyperinsulinemic hypoglycemia: The pathologist's contribution. Histopathology 1998;32:15–19.
13 Junien C: Beckwith-Wiedemann syndrome, tumorigenesis and imprinting. Curr Opin Genet Dev 1992;2:431–438.
14 Giannoukakis N, Deal C, Paquette J, Goodyer C, Polychronakos C: Parental genomic imprinting of the human IGF2 gene. Nat Genet 1993;4:98–101.
15 Lee MP, Hu R-J, Johnson LA, Feinberg AP: Human KVLQT1 gene shows tissue-specific imprinting and encompasses Beckwith-Wiedemann syndrome chromosomal rearrangements. Nat Genet 1997;15:181–185.
16 Matsuoka S, Edwards M, Bai C, Parker S, Zhang P, Baldini A, Harper J, Elledge S: p57KIP2, a structurally distinct member of the p21CIP1 Cdk inhibitor family, is a candidate tumor suppressor gene. Genes Dev 1995;9:650–662.
17 Zhang Y, Tycko B: Monoallelic expression of the human H19 gene. Nat Genet 1992;1:40–44.
18 Hao Y, Crenshaw T, Moulton T, Newcomb E, Tycko B: Tumour-suppressor activity of H19 RNA. Nature 1993;365:764–767.
19 Hatada I, Inazawa J, Abe T, Nakayama M, Kaneko Y, Jinno Y, Niikawa N, Ohashi S, Fukushima Y, Iida K, Yutani C, Takahashi S, Chiba Y, Ohishi S, Mukai T: Genomic imprinting of human p57KIP2 and its reduced expression in Wilms' tumors. Hum Mol Genet 1996;5:783–788.
20 Guillemot F, Caspary T, Tilghman SM, Copeland NG, Gilbert DJ, Jenkins NA, Anderson DJ, Joyner AL, Rossant J, Nagy A: Genomic imprinting of Mash2, a mouse gene required for trophoblast development. Nat Genet 1995;9:235–242.
21 Ohlsson R, Nystrom A, Pfeifer-Ohlsson S, Tohonen V, Hedborg F, Schofield P, Flam F, Ekström T: IGF2 is parentally imprinted during human embryogenesis and in the Beckwith-Wiedemann syndrome. Nat Genet 1993;4:94–97.
22 Yuan L, Qian N, Tycko B: An extended region of biallelic gene expression and rodent-human synteny downstream of the imprinted H19 gene on chromosome 11p15.5. Hum Mol Genet 1996;5:1931–1937.
23 Steenman M, Rainier S, Dobry C, Grundy P, Horon I, Feinberg A: Loss of imprinting of IGF2 is linked to reduced expression and abnormal methylation of H19 in Wilm's tumour. Nat Genet 1994;7:433–439.
24 Dubois J, Brunnelle F, Touati G, Sebag G, Nuttin C, Thach T, Nihoul-Fékété C, Rahier J, Saudubray J: Hyperinsulinism in children: Diagnostic value of pancreatic venous sampling correlated with clinical, pathological and surgical outcome in 25 cases. Pediatr Radiol 1995;25:512–516.
25 Aguilar-Bryan L, Nichols C, Wechsler S, Clemen TJ, Boyd A, Gonzalez G, Herrera Sosa H, Nguy K, Bryan J, Nelson D: Cloning of the beta cell high-affinity sulfonylurea receptor: A regulator of insulin secretion. Science 1995;268:423–426.
26 Inagaki N, Inazawa J, Seino S: cDNA sequence, gene structure, and chromosomal localization of the human ATP sensitive potassium channel, uKATP-1, gene. Genomics 1995;30:102–104.
27 Kane C, Shepherd RM, Squires PE, Johnson PRV, James RFL, Milla PJ, Aynsley-Green A, Lindley KJ, Dunne MJ: Loss of functional K_{ATP} channels in pancreatic β-cells causes persistent hyperinsulinemic hypoglycemia of infancy. Nat Med 1996;2:1344–1347.
28 Reik W, Brown KW, Schneid H, Le Bouc Y, Bickmore W, Maher ER: Imprinting mutations in the Beckwith-Wiedemann syndrome suggested by altered imprinting pattern in the IGF2-H19 domain. Hum Mol Genet 1995;4:2379–2385.
29 Rahier J: Relevance of endocrine pancreas nesidioblastosis to hyperinsulinemic hypoglycemia. Diabetes Care 1989;12:164–166.
30 Ledbetter DH, Engel E: Uniparental disomy in humans: Development of an imprinting map and its implications for prenatal diagnosis. Hum Mol Genet 1995;4:1757–1764.
31 Sempoux C, Poggi F, Brunelle F, Saudubray J, Fékété C, Rahier J: Nesidioblastosis and persistent neonatal hyperinsulinism. Diabète Métab 1995;21:402–407.
32 Brasier J, Henske E: Loss of the polycystic kidney disease (PKD1) region of chromosome 16p13 in renal cyst cells supports a loss-of-function model for cyst pathogenesis. J Clin Invest 1997;99:194–199.
33 Lyonnet S, Bonnefont J-P, Saudubray J-M, Nihoul-Fékété C, Brunelle F: Localization of focal lesion permitting partial pancreatectomy in infants. Lancet 1989;ii:671.

Hormone Research

Horm Res 2000;53(suppl 1):7–11

Neonatal Diabetes: New Insights into Aetiology and Implications

Julian P.H. Shield
on behalf of the Transient Neonatal Diabetes Mellitus Research Group[1]

Institute of Child Health, Bristol, UK

Key Words
Transient neonatal diabetes mellitus · Permanent neonatal diabetes mellitus · Hyperglycaemia · Type 1 diabetes mellitus · Type 2 diabetes mellitus

Abstract
Neonatal diabetes mellitus (NDM) is defined as hyperglycaemia occurring in the first few weeks of life. It can be either transient (TNDM) or permanent (PNDM), and until recently, little was known about the condition. A cohort of 30 infants with a history of TNDM has been studied, and findings have suggested that NDM does not have the same aetiology as classical type 1 childhood diabetes. Uniparental isodisomy of chromosome 6 and an unbalanced duplication of paternal chromosome 6 have both been described as a genetic basis for TNDM in over 75% of the cases. In addition, cerebellar hypoplasia and Walcott-Rallison syndrome have been associated with PNDM, suggesting an autosomal recessive inheritance pattern; furthermore, a mutation in the gene insulin promoter factor 1 has been identified as a cause of pancreatic agenesis in PNDM. In the long term, TNDM may reduce beta cell functional capacity and present a predisposition to type 2 diabetes mellitus.

Introduction

Until recently, little was known about the unusual clinical manifestation of insulin-dependent diabetes mellitus arising in the neonatal period, though its first description as a clinical entity was by Kitselle [1] in 1852 who described it in his own son. Classical neonatal diabetes mellitus (NDM) is defined as hyperglycaemia occurring in the first 6 weeks of life in term infants. Clinically, the condition can be divided simply on the grounds of resolution or persistence past the 1st year of life. Transient NDM (TNDM) has an incidence of around 1:400,000 live births, with permanent NDM (PNDM) developing in the neonatal period seemingly rarer [2, 3]. Our group has now recruited a cohort of 30 infants with a history of TNDM and has studied it extensively. Much of the data presented in this article directly relate to the findings from this cohort.

[1] J.P.H. Shield, I.K. Temple, R. Gardner, D.O. Robinson, D.J.G. MacKay and Wessex Regional Genetics Service and Laboratory.

KARGER
Fax + 41 61 306 12 34
E-Mail karger@karger.ch
www.karger.com

0301–0163/00/0537–0007$17.50/0

Accessible online at:
www.karger.com/journals/hre

Dr. J.P.H. Shield
Institute of Child Health
St. Michael's Hill
Bristol BS2 8BJ (UK)
Tel. +44 117 9215411, Fax +44 117 9285389, E-Mail j.p.h.shield@bristol.ac.uk

Clinical Characteristics

In TNDM, infants usually develop hyperglycaemia within the 1st week of life. The median age of presentation within our TNDM cohort was day 3 of life, and there was equal sex distribution. The majority of the infants were born small for gestational age with a median birth weight of 2.1 kg below the 0.4th percentile (range 1.60–3.37 kg). Marked hyperglycaemia occurs in the face of minimal ketosis [2–6], with peak blood sugar levels remarkably high, at times greater than 70 mmol/l [2]. In our cohort 11 infants (37%) had insulin or 'C' peptide levels estimated at the time of initial diagnosis, and universally these have shown absent or minimal insulin responses in the face of severe hyperglycaemia, suggesting the primary condition is due to inadequate production and/or secretion of insulin. The median requirement for exogenous insulin therapy was 12 weeks, but the longest recorded in our group was 14 months.

In 13 (43%) of the infants, islet cell antibodies were assessed at diagnosis and all proved negative. In 1996, Mazzella et al. [7] suggested that DQB1 HLA typing of children with NDM could predict those liable to go on to PNDM, implying an auto-immune aetiology. However, we found that children who eventually proved to have PNDM could carry diabetes-resistant alleles (aspartate position 57, β-chain positive), while those with TNDM could equally carry classical susceptibility alleles without islet cell positivity [8]. This combination of findings suggests that the spectrum of neonatal diabetes is not related to the auto-immune phenomena classically responsible for childhood diabetes.

Neonatal Diabetes Syndromes

There have been a number of reported associations with TNDM: both umbilical herniae [9] and macroglossia [10] have been described. In our cohort of infants, 23% manifested macroglossia, and 7% had umbilical herniae. These two anomalies are interesting given their association with Beckwith-Wiedemann syndrome in which there is typically hyperinsulinaemia [11]. Some of the original descriptions of TNDM indicated a relatively poor prognosis for developmental outcome, but analysis of these reports reveals episodes of significant hypoglycaemia during the initial phases of treatment in these children [12]. The two largest follow-up studies have found little evidence of adverse development [2, 3], suggesting that the initial reports were strongly influenced by problems of therapy with recurrent hypoglycaemia rather than primary abnormalities in development associated with the disease per se.

In PNDM a number of syndromic constellations have been described. Most recently, three members of a consanguineous family have been described with neonatal diabetes and cerebellar hypoplasia with a suggested autosomal recessive inheritance pattern. The infants all died within a few months after birth from a combination of metabolic dysfunction, respiratory compromise and apparent sepsis [13]. Interestingly, there are a number of specific transcriptional activators that regulate gene expression which beta cells and neuronal tissues have in common [14]; dysfunction in any of these might explain both components of this syndrome. Wolcott-Rallison is another neonatal diabetes syndrome with apparent autosomal recessive inheritance. Infants develop permanent diabetes in the first few weeks of life associated with a multiple epiphyseal dysplasia. Additional features include seizures (not attributable to hypoglycaemia), developmental delay, renal insufficiency, hepatomegaly and early death [15, 16].

Two further syndromes have been described, with PNDM as a major component inherited in a presumed X-linked fashion. There have now been two reports of families with associated intractable diarrhoea, immune dysregulation, erythroderma in some cases and severe failure to thrive [17, 18]. Necropsy in 1 case identified agenesis of the islets of Langerhans [17]. Phosphoribosylpyrophosphate synthetase superactivity, a rare X-linked condition, has also been described in two brothers born small for gestational age with PNDM and hyperuricaemia. They developed a progressive axonal neuropathy in association with growth and mental retardation and ataxia. The mother developed gestational diabetes with hypertonia and hyperuricaemia [19].

PNDM has been described with both complete pancreatic agenesis and pancreatic hypoplasia associated with significant exocrine dysfunction [20, 21], while in other cases of PNDM there have been no exocrine insufficiency and evidence of minimal but significant residual islet cell function [22].

Genetics

Transient NDM

We can now divide the genetic basis of this condition into three distinct groups. The first anomaly described in TNDM was uniparental isodisomy of chromosome 6

(UPD 6) in which two haplo-identical copies of chromosome 6 are inherited from the father with no contribution from the mother [23]. A number of further reports [24–26] have confirmed the above association, while the single case of maternal UPD 6 did not develop TNDM. In our cohort, we have identified UPD 6 in over a third of individuals (38%). The risk of TNDM in further parental pregnancies is minimal.

The second basis for TNDM is an unbalanced duplication of paternal chromosome 6 in the region 6q24 (long-arm chromosome 6) [27]. With this anomaly the offspring inherit a small extra segment of paternal chromosome 6, which makes them liable to NDM. Only if this duplication comes from the father does inheritance result in diabetes, suggesting that the gene(s) is/are imprinted. With imprinted genes, the parent of origin defines the phenotypic expression. These small duplications, which are not all of similar length, have allowed us to restrict the area of interest containing the gene responsible for the condition [28]; duplications defining a similar area of interest have now been described [29]. Within our cohort these duplications are responsible for diabetes in just over a third of cases (38%).

Third, in nearly a quarter (24%) of our infants we have so far been unable to elucidate an anomaly in chromosome 6 to account for the condition. We have found that the developmental anomalies associated with TNDM are not associated with any particular genetic anomaly.

Permanent Neonatal Diabetes mellitus

The first specific gene identified as a cause of NDM was described by Stoffers et al. [30] in 1997. Insulin promoter factor 1 (IPF-1) is a homeodomain transcription factor required for the development of the pancreas and also for glucose-responsive stimulation of insulin gene transcription [31]. In a child with pancreatic agenesis the investigators identified an inactivating cytosine deletion in the protein coding sequence of the gene IPF-1, for which the child was homo-allelic [30]. Furthermore, the family of this child contained a number of members who developed early-onset type 2 diabetes and were identified as being hetero-allelic for this mutation. This single mutation actually results in the translation of two isoforms of mutated IPF-1, one of which inhibits the function of wild-type IPF-1, thereby exhibiting a dominant negative effect in the hetero-allelic state [32].

Other pancreatic developmental genes have also been studied in relation to PNDM. In 1997, investigators demonstrated that *Pax* 4 is essential for the differentiation, in early development, of endocrine progenitors into insulin-producing beta cells in mammalian pancreas. Mutant mice homozygous for a deletion in *Pax* 4 produced no beta cells, but an excess of alpha cells, probably by way of a default pathway [33]. As this was felt to be somewhat analogous to the histopathology of the pancreas in Wolcott-Rallison syndrome [34], the gene was examined as a candidate, but has now been excluded [35].

Long-Term Outcome of TNDM

It was originally reported that TNDM carries no long-term sequelae [36]. However, a number of reports have now established that relapse usually in later childhood is common [37–40]. Furthermore, there is evidence that, although overt diabetes remits early in life, residual pancreatic function is suboptimal, as shown by reports of transient hyperglycaemia throughout childhood at times of stress, such as during intercurrent illnesses [2]. As might be expected, relapse when it occurs, appears not to have an auto-immune basis [40], and patients characteristically require far less insulin to maintain good control than those with classical type 1 diabetes [2]. In our cohort, 11 of the 18 (61%) children who were over 4 years of age experienced relapse of diabetes; the mean age of recurrence was 14 (range 4–25) years. Eight of these patients required exogenous insulin therapy, though in 2 the requirement has been intermittent. Others have been managed with diet alone or oral hypoglycaemic agents. The genetic mechanism identified as the primary cause of TNDM has not influenced the outcome in terms of relapse rate later in life.

Discussion

TNDM is a rare condition manifesting as intra-uterine growth retardation and deficiency in endogenous insulin secretion in early life which resolves usually within 3–4 months after birth. While in the more seriously affected individuals the condition develops to manifest clinical disease, it is likely that in others clinical illness does not occur, the individuals only developing mild, subclinical hyperglycaemia [41]. Therefore, the recorded incidence figures are probably an underestimation. Whatever the neonatal outcome, there appears to be residual pancreatic dysfunction or reduced capacity allowing the development of type 2 diabetes in later adolescence or early adult life. TNDM and at least some forms of PNDM have a genetic basis. The IPF-1-inactivating mutation

(Pro63fsdelC) is also of interest, as it favours the development of type 2 diabetes in the heterozygous state due to its function as a pancreatic development gene. In the 'Barker-Hales' hypothesis, adverse intra-uterine environmental events predict poor fetal growth and late-onset type 2 diabetes [42]. TNDM was one of the first conditions in which a genetic anomaly predicted both severe intra-uterine growth retardation and a propensity to later diabetes. This suggests that the spectrum of neonatal diabetes encompasses a number of conditions related to defects in pancreatic developmental genes. In the case of TNDM, the gene(s) additionally would appear to be imprinted because only inheritance of chromosome 6 anomalies through the paternal side results in phenotypic expression. This extends the evidence that the gene is developmental and important in fetal growth because imprinting mechanisms appear central to the control of intra-uterine growth [43]. Further genetic predictors of birth weight have now been described, including the variable number tandem repeat sequence in the insulin gene region [44] and the MODY 2 gene [45]. Taken as a whole, there is now much evidence to suggest that birth weight and risk of type 2 diabetes in later life are at least partly defined by genetic inheritance patterns and not solely by environmental variables.

Conclusions

Identification of a gene involved in TNDM will allow evaluation of its role in pancreatic development. Given the association of TNDM and intra-uterine growth retardation and the apparent effect of reducing beta cell functional capacity, and the linked predisposition to type 2 diabetes mellitus, further evaluation of this gene will include potential links with the development of type 2 diabetes in the general population.

References

1 Kitselle JFJ: Kinderheilkunde 1852;18:313.
2 Shield JPH, Gardner RJ, Wadsworth EJK, James RS, Robinson DO, Baum JD, Temple IK: Neonatal diabetes: A study of its aetiopathology and genetic basis. Arch Dis Child 1997; 76:F39–F42.
3 von Mühlendahl KE, Herkenhoff H: Long-term course of neonatal diabetes. N Engl J Med 1995;333:704–708
4 Gentz JCH, Cornblath M: Transient diabetes of the newborn. Adv Pediatr 1969;16:345–363.
5 Ferguson IC: Neonatal hyperglycaemia: Case report with plasma insulin studies. Arch Dis Child 1967;42:509–513.
6 Kuna P, Addy DP: Transient neonatal diabetes mellitus. Treatment with chlorpropamide. Am J Dis Child 1979;133:65–66.
7 Mazzella M, Cotellessa M, Bonacci W, Bruno C, Serra G: Neonatal diabetes and DQB1 typing. Diabetes Care 1996;19:1308.
8 Shield JPH, Howell WM, Temple IK: Neonatal diabetes. Diabetes Care 1997;20:1045–1046.
9 Salerno MC, Gasparini N, Sandomenico ML, Franzese A, Tenore A: Two interesting cases of transient neonatal diabetes mellitus. J Pediatr Endocrinol 1994;7:47–52.
10 Dacou-Voutetakis C, Anagnostakis D, Xanthou M: Macroglossia, transient neonatal diabetes mellitus and intrauterine growth failure: A new distinct entity? Pediatrics 1975;55:127–131.
11 Weng EY, Moeschler JB, Graham JM Jr: Longitudinal observations on 15 children with Wiedemann-Beckwith syndrome. Am J Med Genet 1995;56:366–373.
12 Hutchinson JH, Keay AJ, Kerr MM: Congenital temporary diabetes mellitus. Br Med J 1962;ii:436–440.
13 Hoveyda N, Shield JPH, Garrett C, Chong WK, Beardsall K, Bentsi-Enchill E, Mallya H, Thompson MH: Neonatal diabetes mellitus and cerebellar hypoplasia/agenesis: Report of an association in three members of a consanguineous family. J Med Genet 1999;36:700–704.
14 Atouf F, Czernichow P, Scharfmann R: Expression of neuronal traits in pancreatic beta cells. J Biol Chem 1997;272:1929–1934.
15 Wolcott CD, Rallison ML: Infancy-onset diabetes mellitus and multiple epiphyseal dysplasia. J Pediatr 1972;80:292–297.
16 Al-Gazali LI, Makia S, Azzam A, Hall CM: Wolcott-Rallison syndrome. Clin Dysmorphol 1995;4:227–233.
17 Peake JE, McCrossin RB, Byrne G, Shepherd R: X-linked immune dysregulation, neonatal insulin dependent diabetes, and intractable diarrhoea. Arch Dis Child 1996;74:F195–199.
18 Jonas M, Bell M, Eidson M, Koutouby R, Hensley G: Congenital diabetes mellitus and fatal secretory diarrhoea in two infants. J Pediatr Gastroenterol Nutr 1991:13;415–425.
19 Christen HJ, Hanefield F, Duley JA, Simmonds HA: Distinct neurological syndrome in two brothers with hyperuricaemia. Lancet 1992;340:1167–1168.
20 Winter WE, Maclaren NK, Riley WJ, Toskes PP, Andres J, Rosenbloom AL: Congenital pancreatic hypoplasia: A syndrome of exocrine pancreatic and endocrine insufficiency. J Pediatr 1986;109:465–468.
21 Wright NM, Metzger DL, Borowitz SM, Clarke WL: Permanent neonatal diabetes mellitus and pancreatic exocrine insufficiency resulting from congenital pancreatic agenesis. Am J Dis Child 1993;147:607–609.
22 Knip M, Koivisto M, Kaar M-L, Puukka R, Kouvalainen K: Pancreatic islet cell function and metabolic control in an infant with permanent neonatal diabetes. Acta Paediatr Scand 1983;72:303–307.
23 Temple IK, James RS, Crolla JA, Sitch FL, Jacobs PA, Howell WM, Betts P, Baum JD, Shield JP: An imprinted gene(s) for diabetes? Nat Genet 1995;9:110–112.
24 Gardner RJ, Robinson DO, Lamont L, Shield JPH, Temple IK: Paternal uniparental disomy of chromosome 6 and transient neonatal diabetes mellitus. Clin Genet 1998;54:522–525.
25 Christian SL, Rich BH, Loebl C, Israel J, Vasa R, Kittikamron K, Spiro R, Rosenfield R, Ledbetter DH: Significance of genetic testing for paternal uniparental disomy of chromosome 6 in neonatal diabetes mellitus. J Pediatr 1999; 134:42–46.
26 Whiteford ML, Narendra A, White MP, Cooke A, Wilkinson AG, Robertson KJ, Tolmie JL: Paternal uniparental disomy for chromosome 6 causes transient neonatal diabetes. J Med Genet 1997;34:167–168.
27 Temple IK, Gardner RJ, Robinson DO, Kibirige MS, Ferguson AW, Baum JD, Barber JC, James RS, Shield JP: Further evidence for an imprinted gene for neonatal diabetes localised to chromosome 6q22–23. Hum Mol Genet 1996;5:1117–1121.

28 Gardner RJ, Mungall AJ, Dunham I, Barber JCK, Shield JPH, Temple IK, Robinson DO: Localisation of a gene for transient neonatal diabetes mellitus to an 18.72cR$_{3000}$ interval on chromosome 6q. J Med Genet 1999;36:192–196.
29 Pivnik EK, Qumsieyeh MB, Tharapel AT, Summitt JB, Wilroy RS: Partial duplication of the long arm of chromosome 6: A clinically recognisable syndrome. J Med Genet 1990;27:523–526.
30 Stoffers DA, Zinkin NT, Stanojevic V, Clarke WL, Habener JF: Pancreatic agenesis attributable to a single nucleotide deletion in human IPF-1 coding region. Nat Genet 1997;15:106–110.
31 Jonsson JL, Carlsson T, Edlund T, Edlund H: Insulin-promoter-factor 1 is required for pancreas development in mice. Nature 1994;371:606–609.
32 Stoffers DA, Stanojevic V, Habener JF: Insulin promoter factor-1 gene mutation linked to early onset type 2 diabetes mellitus directs expression of dominant negative isoprotein. J Clin Invest 1998;102:232–241.
33 Sosa-Pineda B, Chowdhury K, Torres M, Oliver G, Gruss P: The Pax 4 gene is essential for differentiation of insulin-producing β cells in the mammalian pancreas. Nature 1997;386:399–402.
34 Stoss H, Pesch H-J, Pontz B, Otten A, Spranger J: Wolcott-Rallison syndrome, diabetes mellitus and spondyloepiphyseal dysplasia. Eur J Pediatr 1982;138:120–129.
35 Bonthron DT, Dunlop N, Barr DGD, El Sanousi AA, Al-Gazali LI: Organisation of the human pax 4 gene and its exclusion as a candidate for the Wolcott-Rallison syndrome. J Med Genet 1998;35:288–292.
36 Shield JPH, Baum JD: Is transient neonatal diabetes a risk factor for diabetes in later life? Lancet 1993;341:693.
37 Shield JPH, Baum JD: Transient neonatal diabetes and later onset diabetes: A case of inherited insulin resistance. Arch Dis Child 1995;72:56–57.
38 Weimerskirch D, Klein DJ: Recurrence of insulin-dependent diabetes mellitus after transient neonatal diabetes: A report of two cases. J Pediatr 1993;122:598–600.
39 Vanelli M, De Fanti A, Cantoni S, Chiari G: Transient neonatal diabetes mellitus: A relapse after 10 years of complete remission. Acta Diabetol 1994;31:116–118.
40 Gottschalk ME, Schatz DA, Clare-Salzler M, Kaufman DL, Ting GSP, Geffner ME: Permanent diabetes without serological evidence of autoimmunity after transient neonatal diabetes. Diabetes Care 1992;15:1273–1276.
41 McGill JJ, Roberton DM: A new type of transient neonatal diabetes mellitus of infancy? Arch Dis Child 1986;61:334–336.
42 Hales CN, Barker DJP, Clark PMS, Cox LJ, Fall C, Osmond C, Winter PD: Fetal and infant growth and impaired glucose tolerance at age 64. BMJ 1991;303:1019–1022.
43 Barlow DP: Genetic imprinting in mammals. Science 1995;270:1610–1613.
44 Dunger DB, Ong KKL, Huxtable SJ, Sherriff A, Woods KA, Ahmed ML, Golding J, Pembrey ME, Ring S, Bennett ST, Todd JA: Association of the INS-VNTR with size at birth. ALSPAC Study Team. Nat Genet 1998;19:98–100.
45 Hattersley AT, Beards F, Ballantyne E, Appleton M, Harvey R, Ellard S: Mutations in the glucokinase gene of the fetus resulting in reduced birth weight. Nat Genet 1998;19:268–270.

Hormone Research

Horm Res 2000;53(suppl 1):12–18

Molecular Pathogenesis of Neonatal Hypothyroidism

Heiko Krude Heike Biebermann Dirk Schnabel Petra Ambrugger Annette Grüters

Department of Pediatrics, Charité University Hospital, Humboldt University, Berlin, Germany

Key Words
Congenital hypothyroidism · Thyroid dysgenesis · Pathogenesis, hypothyroidism

Abstract
In patients with congenital hypothyroidism (CH), the autosomal recessive inheritance of mutations of thyroid peroxidase, thyroglobulin and the NIS and pendrin genes encoding for sodium iodide transporters has been identified. CH due to thyroid dysgenesis was considered to be a sporadic disease, but recently, inheritable defects of thyroid development have been described. The autosomal recessive inheritance of mutations of the thyroid-stimulating hormone receptor gene was recognized in patients with CH and thyroid hypoplasia, while autosomal dominant mutations of the Pax-8 gene were described in patients with thyroid dysgenesis. In addition, analysis of mutations of the β-thyrotropin gene has resulted in a new understanding of the pathogenesis of central CH. Molecular genetic studies in patients with CH detected by newborn screening will provide the information necessary for genetic counselling and may help to explain the less favourable outcome present in 5–10% of the patients.

Introduction

Congenital hypothyroidism (CH) is the most frequent congenital endocrine disorder, occurring with a consistent frequency of 1 in 3,000–4,000 newborns. In only very few patients has the defect of thyroid hormone biosynthesis been explained by the autosomal recessive inheritance of mutations of the thyroid peroxidase (TPO) gene [1–4] or the thyroglobulin (TG) gene [5]. More recently, mutations of the sodium iodide symporter (NIS) and the pendrin gene, also encoding for an iodide transporter, explained the autosomal recessive inheritance of CH with goitre development [6, 7]. The spectrum of developmental defects (athyrosis, hypoplasia, ectopy) of the thyroid gland, summarized as thyroid dysgenesis, which account for more than 80% of all cases of CH, was not thought to have been caused by inherited gene defects because of their sporadic occurrence. However, some families with a multiple occurrence of thyroid dysgenesis have been described with a pattern of inheritance compatible with an autosomal dominant trait of inheritance. In addition, a striking preponderance in female patients with thyroid dysgenesis, but not in patients with defects of hormone biosynthesis, has been noticed.

Controversial discussions have focused on whether mutations of genes for thyroid-specific proteins – (e.g., thyroid-stimulating hormone (TSH)-receptor gene – are

KARGER
Fax + 41 61 306 12 34
E-Mail karger@karger.ch
www.karger.com

0301–0163/00/0537–0012$17.50/0

Accessible online at:
www.karger.com/journals/hre

Prof. A. Grüters
Department of Pediatrics, Charité University Hospital
Campus Virchow, Augustenburger Platz 1
D–13353 Berlin (Germany)
Tel. +49 30 45066251, Fax +49 30 45066926, E-Mail annette.grueters@charite.de

involved in thyroid growth and differentiation. Moreover, transcription factors taking part in the induction and migration of the thyroid gland may result in CH. Whether the different types of thyroid dysgnesis are caused by mutations in different genes or whether a phenotypical spectrum ranging from athyrosis to hypoplasia can be caused by the same gene defect are details that remain to be elucidated. Thus, screening for mutations in known candidate genes was initiated by several groups of investigators [8–10]. This report summarizes the current knowledge on the molecular pathogenesis of CH.

Table 1. Summary of the TPO mutations detected in patients with CH

Exon	Position	Nucleotide change	Mutation
2	141	20 bp duplication	frame shift
8	1277	GGCC duplication	frame shift
9	1429	A→T	Ile→Phe
9	1447	T→G	Tyr→Asp
10	1708	C→T	Arg→stop
10	1858	G→A	Gly→Ser
11	2068	C→G	Gln→Glu
11	2033	G→A	Arg→Gln
14	2485	G→A	Glu→Lys
14	2505	C insertion	frame shift

Primary CH

Mutations of the TPO Gene

Inborn errors of thyroid hormone biosynthesis have been postulated in newborns with hypothyroidism and a normally developed thyroid gland, but to date only in isolated cases have these defects been shown to be caused by the inheritance of mutated genes. The most comprehensive search for mutations by denaturating gradient gel electrophoresis was made by de Vijlder and co-workers in Dutch patients with CH [1–3]. They described numerous polymorphisms of the TPO gene and were able to prove by functional studies that some of them will cause total or partial organification defects resulting in CH (table 1). We performed a screening of the TPO gene for mutations in relevant functional regions: 23 patients with CH and normally developed glands and 3 patients with a congenital goitre were screened by single-stranded conformational polymorphism and restriction enzyme analysis. We detected 1 patient of non-consanguineous parents of German origin to be homozygous for the same mutation (CCGG) as that which had been previously described in a Brazilian and in a Dutch patient [4]. Surprisingly, the Dutch patients with total organification defects identified during the newborn period also lacked a thyroid enlargement; however, a goitre has been regarded as a hallmark of organification defects in patients diagnosed during later life.

Mutations of the TG Gene

To date only five mutations of the TG gene have been described. These patients were characterized by a familial history of CH and goitre development [5]. From these studies, it is not apparent whether the goitre was already present at birth. The analysis of immunoreactive TG revealed both unmeasurable and measurable TG serum levels. These results are most likely caused by differences of the TG molecule, leakage from thyroid tissue and the immunoassays used. Therefore, the clinical and biochemical phenotypes of patients with CH due to mutations of the TG gene are not well defined, and no systematic search for mutations by newborn screening has been performed. As the TG gene has a size of 300 kb, and because the coding region encompasses 37 exons, mutation screening can be justified only when more practical and less time-consuming methods are available.

Mutations of the NIS and Pendrin Genes

Two patients have been described in whom CH was caused by homozygosity for either a missense mutation (T354P) in the ninth segment of the transmembrane domain [10] or a non-sense mutation (C272X) in the fourth extracellular loop [7] of the NIS. Both mutants resulted in a complete lack of iodide transport activity in transfection experiments. A 3rd patient with CH was compound heterozygous for two missense mutations (Q267E and nucleotide 1940) leading to a NIS molecule lacking 129 carboxy-terminal amino acids [7]. In all 3 patients the thyroid gland could not be visualized by isotope scanning, though a goitre developed in early life. However, similar to the patients with organification defects, goitre development did not start prenatally.

Patients with missense or non-sense mutations of the pendrin gene [11] present with a large phenotypical spectrum ranging from euthyroidism without thyroid enlargement to the full clinical picture of congenital goitre and hypothyroidism with hereditary hearing loss [6, 12]. The pendrin gene, which was first thought to encode for a putative sulphate transporter, also encodes for a chloride-iodide transporter [13]. Therefore, iodide is incompletely organified, and thyroid hormone biosynthesis is variably affected.

The molecular basis of hearing loss is still unknown. In Berlin, two siblings of consanguineous parents with CH detected by neonatal screening and with congenital goitre and hearing loss were found to have a homozygous mutation (R409H) of the pendrin gene [6]. Despite early and adequate treatment with iodine-thyroxine, developmental delay was observed in both children. Although the analysis of the pendrin gene demonstrates a molecular heterogeneity for mutations, the molecular analysis of the pendrin gene in the assessment of patients with CH and hearing loss with or without goitre is encouraged, because the results will be important for the understanding of the phenotypical spectrum and biological importance of the iodide transporter in other tissues besides the thyroid gland.

Loss-of-Function Mutations of the TSH Receptor Gene

Based on in vitro studies of thyroid tissue of patients with CH, resistance to TSH was postulated as a possible pathogenetic mechanism [14, 15]. With the availability of molecular genetic analysis, compound heterozygosity and homozygosity for missense and non-sense mutations were described in patients with moderate [8] and severe [9] CH. Although some patients were considered to have athyrosis due to the absence of visible thyroid tissue on thyroid ^{99}Tc or ^{123}I scans, subsequent ultrasound studies have revealed hypoplastic thyroid remnants in all of them. Moreover, thyroglobulin was clearly measurable or even elevated in these patients with 'apparent athyrosis' [16]. While some of the non-sense mutations severely alter receptor function (e.g. I167L) [17], other mutations are not expressed in the plasma membrane, but are retained within the cell, as demonstrated by expression studies [8]. Some missense mutations of the coding region (e.g., R109E) have been shown to only mildly decrease receptor function [18]. Most of the more severe loss-of-function mutations of the TSH receptor that have been described to date are located in the transmembrane domain or very close to the transmembrane part of the receptor, and, very recently, an intronic mutation in the critical splice region (–1, intron 5) was found to cosegregate with CH in a consanguineous family (fig. 1).

Mutations of Transcription Factors (TTF-1, TTF-2, Pax-8)

Studies in mice have demonstrated that several transcription factors (TTF-1, TTF-2 and Pax-8) are expressed transiently or permanently in the developing thyroid gland and govern its normal induction, migration and differentiation [19, 20]. After cloning of the human genes, a screening for mutations in patients with CH was initiated at first for TTF-1. However, extensive screening by single-stranded conformational polymorphism and direct sequencing in patients with CH did not reveal mutations in a large cohort of Italian patients with CH [21]. Recently, 1 patient was described with a deletion of chromosome 14 q 13-21, including the locus of the TTF-1 gene on one allele [22]. However, hypothyroidism was only mild with a TSH level of 48 mU/l and a normal T_4 concentration of 13.9 μg/dl. Interestingly, the history of this patient was complicated by a severe respiratory distress syndrome requiring ventilation for 1 week. Moreover, at the age of 2 years, mental and motor development was delayed despite thyroxine replacement which began on day 19 of life. It was suggested that haplo-insufficiency of TTF-1 was responsible for mild hypothyroidism and neonatal respiratory distress, because TTF-1 is expressed not only in the developing thyroid, but also in the lung leading to severe respiratory distress in the TTF-1 knockout mice. Although TTF-1 is also expressed in ventral forebrain and pituitary gland [19], it remains speculative as to whether TTF-1 deficiency contributes to the developmental delay, because other genes in the deleted region may also be involved. To date, this is the only known case of TTF-1 deficiency in humans.

TTF-2 is a second transcription factor expressed during the embryonic development of the thyroid gland which regulates the expression of thyroid-specific proteins like TPO and TG. Targeted disruption of the TTF-2 gene in mice leads to hypothyroidism due to defective migration ranging from athyrosis to ectopy, but also to cleft palate formation [20]. Two particular cases of CH characterized by spiky hair and cleft palate, who had been described previously [23, 24], were recently shown to be affected by a homozygous missense mutation (A65V). In addition, these two brothers had choanal atresia and a bifid epiglottis [18]. The mutant protein was shown to have impaired DNA binding and transcriptional function. However, the screening for mutations in the human TTF-2 gene also failed to demonstrate their presence in patients with non-syndromic CH [25]. Moreover, we were not able to detect mutations of the coding region of the TTF-2 gene in a female patient with CH due to athyrosis, cleft palate and ventricular septum defect by direct sequencing.

Autosomal dominant Pax-8 mutations in patients with variable degrees of thyroid hypoplasia as well as thyroid ectopy were found [10]. We have detected an autosomal

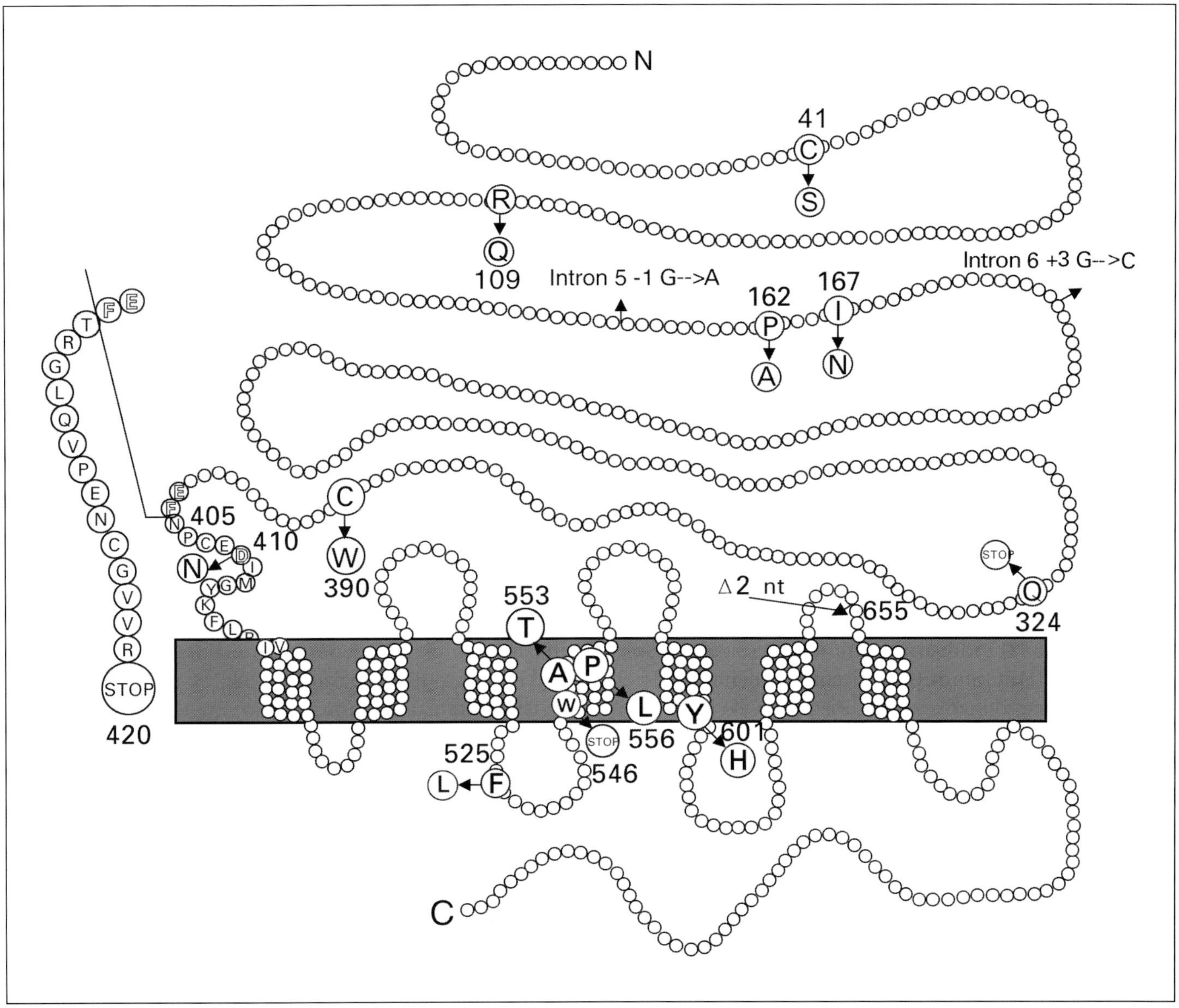

Fig. 1. Inactivating mutations of the TSH receptor gene. The positions of missense and non-sense mutations of the TSH receptor gene are indicated. In patients with congenital hypothyroidism the following mutations are found: C390W, D406-412 and insertion, A553T, Δ2 nt 655, intron 6+3, intron 5-1. All other mutations were found in patients with hyperthyrotropinaemia.

dominant Pax-8 missense mutation [10] in 2 patients with a familial history of hypothyroidism of variable severity. The mother was diagnosed with severe hypothyroidism at the age of 8 years. Her first-born son was detected by newborn screening to have severe CH and thyroid hyoplasia with a cystic degeneration, while in the second-born daughter CH was moderate as compared with her brother, but the thyroid was again hypoplastic. The mutation (A62L), which was present in the mother, son and daughter on one allele, abolishes DNA binding of the mutated Pax-8. In addition, we have detected a heterozygous Pax-8 mutation (R108stop) in a patient with CH and thyroid hemiagenesis whose father carrying the same mutation was thought to be euthyroid. Subsequent investigation revealed, however, that he also had a hemithyroidea and mild hyperthyrotropinaemia. The demonstration of

Table 2. Phenotype in CH caused by transcription factor defects in humans

Transcription factor defect	Thyroid phenotype	Associated malformation
TTF-1	thyroid hypoplasia	respiratory distress, chronic lung disease, developmental delay
TTF-2	thyroid hypoplasia ectopy?	choanal atresia, spiky hair, cleft palate
Pax-8	thyroid hypoplasia ectopy?	malformation of the kidney, cryptorchidism

an autosomal dominant trait of inheritance of thyroid dysgenesis in the presence of a moderate clinically inapparent phenotype in the parental generation is not only compatible with most familial cases of CH with thyroid dysgenesis, but also would explain the apparent sporadic occurrence of most cases of CH due to thyroid dysgenesis. Table 2 summarizes the phenotypic spectrum of the transcription factor gene mutations identified so far.

Mutations of the Thyrotropin Gene

Only rarely do familial combined pituitary hormone insufficiencies (e.g. due to mutations of the transcription factors Pit-1 or Prop-1) [26] cause symptoms of hypothyroidism in the early neonatal period. Recently a patient with isolated central hypothyroidism and mental retardation due to compound heterozygosity for loss-of-function mutations of the thyrotropin-releasing hormone receptor gene was described [27]. In addition, mutations of the thyrotropin gene have been described in patients with only moderate [28] and severe isolated central CH [15]. We have observed 3 newborn infants with classical symptoms of CH (prolonged jaundice, sleepiness, feeding problems, hypotonia) despite normal results in TSH screening. However, the thyroid hormone levels were almost undetectable and sequencing of the thyrotropin β gene disclosed homozygosity for a base pair deletion in codon 105, resulting in a frame shift with the consequence of a premature stop at codon 114. This results in a β-TSH peptide with nine non-homologous amino acids and a lack of the five terminal amino acids. Moreover, the same mutation had been described previously in the offspring of a consanguineous Brazilian kindred with multiple occurrence of central hypothyroidism [15] and in another sporadic case in Germany [29]. Therefore, it is possible that this mutation, which leads to a more severe phenotype than other mutations, defines a hot spot for mutations in the β-TSH gene.

Discussion

Screening for CH has successfully enabled an early thyroid hormone replacement and a normal mental development in most affected children [30]. It might be possible that familial CH may occur more frequently than previously thought, because these patients have a normal reproductive fitness; thus, the study of genetic defects resulting in CH is justified from more than just an academic point of view. Molecular genetic studies of the genes for thyroid-specific proteins (e.g. TPO, TG or iodide transporters) allow the clarification of the cause of a defective hormone production in the 15–20% of the patients with morphologically normal thyroid glands and will enable appropriate genetic counselling. In contrast, the hypothesis that genetic defects cause thyroid dysgenesis is still controversial.

A major argument against a genetic cause is the sporadic occurrence of thyroid dysgenesis. As most cases of thyroid dysgenesis that have been investigated to date do not appear with an inheritance compatible with a classical mendelian trait, somatic mutations of thyroid-specific proteins also have to be considered. In this respect it is notable that CH due to thyroid dysgenesis, in contrast to CH with a normal developed gland, is three times more prevalent in females [31]. As TSH is not only important for thyroid function but also for thyroid cell growth, it was logical to screen for mutations in the TSH receptor gene. In a subset of CH and thyroid hypoplasia, TSH receptor mutations have been reported, but compound heterozygosity or homozygosity for TSH receptor mutations, which result in a complete loss of function, so far has not resulted in athyrosis or ectopy, the most frequent developmental defects leading to CH. Moreover, in other patients with resistance to TSH and normally developed or hypoplastic glands, the coding region of the TSH receptor gene was found to be normal, indicating that additional defects either in receptor expression or signal transduction remain to be discovered. Recently the involvement of transcription factors in the development of the thyroid gland has been studied extensively in mice [32], but the screening for mutations in patients with CH has been rather unsuccessful [10, 25]. The reason might be that these transcription factors are not exclusively expressed in the thyroid. The TTF-1 gene is also expressed in the fetal lung and plays an important role in surfactant production as well as in both the developing forebrain and the pituitary gland [32], and humans with TTF-2 mutations had cleft palate and choanal atresia. Therefore, TTF-1 and TTF-2 will not be involved in the pathogenesis of the usual non-syndromic phenotype of CH.

Considering the fact that a subset of patients with CH has a less favourable outcome of mental development despite early and adequate therapy, it is possible that an unknown genetic defect may affect not only thyroid but also brain development. The fact that autosomal dominant inheritance of Pax-8 mutations may result in a phenotypic spectrum ranging from athyrosis and ectopy to hypoplasia and even hemithyroidea supports the hypotheses that the apparent sporadic occurrence of CH due to thyroid dysgenesis does not exclude a genetic cause and that all types of dysgenesis may even result from defects in one gene. However, to date no genetic defects have been found in patients with athyrosis and ectopy, and it is likely, therefore, that more genes involved in thyroid development and differentiation need to be identified.

CH due to isolated thyrotropin deficiency was thought to be of minor importance because of its rarity and because it was not thought to be correlated with severe symptoms of CH leading to mental retardation. Mutational analysis of the β-thyrotropin gene in affected patients has shown that one particular mutation (C105V) leads to severe and early onset CH [33]. Therefore, in contrast to previous assumptions, the complete loss of TSH action caused by either a defective ligand or loss of function of the receptor can result in severe CH. Therefore, clinical symptoms of CH in newborns and infants should result in an immediate investigation of the thyroid function irrespective of the TSH screening result.

Molecular genetic research in congenital hypothyroidism has already made major contributions to the understanding of the pathogenesis in a subset of patients, but the pathogenesis of athyrosis and ectopy remains a mystery that has to be resolved in the future.

References

1 Bikker H, den Hartog MT, Baas F, Gons MH, Vulsma T, de Vijlder JJ: A 20-basepair duplication in the human thyroid peroxidase gene results in a total iodide organification defect and congenital hypothyroidism. J Clin Endocrinol Metab 1994;79:248–252.
2 Bikker H, Vulsma T, Baas F, de Vijlder JJ: Identification of five novel inactivating mutations in the human thyroid peroxidase gene by denaturing gradient gel electrophoresis. Hum Mutat 1995;6:9–16.
3 Bikker H, Waelkens JJ, Bravenboer B, de Vijlder JJ: Congenital hypothyroidism caused by a premature termination signal in exon 10 of the human thyroid peroxidase gene. J Clin Endocrinol Metab 1996;81:2076–2079.
4 Grüters A, Köhler B, Wolf A, Söling A, de Vijlder L, Krude H, Biebermann H: Screening for mutations of the human thyroid peroxidase gene in patients with congenital hypothyroidism. Exp Clin Endocrinol Diabetes 1996;104 (suppl 4):121–123.
5 Ieri T, Cochaux P, Tagorvnik HM, Suzuki M, Shimoda SI, Pereet J, Vassart G: A 3′ splice mutation in the thyroglobulin gene responsible for congenital goiter with hypothyroidism. J Clin Invest 1991;88:1901–1905.
6 Coyle B, Reardon W, Herbrick JA, Tsui LC, Gausden E, Lee J, Coffey R, Grüters A, Grossman A, Phelps PD, Luxon L, Kendall-Taylor P, Scherer SW, Trembath RC: Molecular analysis of the PDS gene in Pendred syndrome. Hum Mol Genet 1998;7:1105–1112.
7 Pohlenz J, Rosenthal IM, Weiss RE, Jhiang SM, Burant C, Refetoff S: Congenital hypothyroidism due to mutations in the sodium/iodide symporter: Identification of a nonsense mutation producing a downstream cryptic 3′ splice site. J Clin Invest 1998;101:1028–1035.
8 Biebermann H, Schoneberg T, Krude H, Schultz G, Gudermann T, Grüters A: Mutations of the human thyrotropin receptor gene causing thyroid hypoplasia and persistent congenital hypothyroidism. J Clin Endocrinol Metab 1997;82:3471–3480.
9 Abramowicz MJ, Duprez L, Parma J, Vassart G, Heinrichs C: Familial congenital hypothyroidism due to inactivating mutation of the thyrotropin receptor causing profound hypoplasia of the thyroid gland. J Clin Invest 1997; 99:3018–3024.
10 Macchia PE, Lapi P, Krude H, Pirro MT, Missero C, Chiovato L, Souabni A, Baserga M, Tassi V, Pinchera A, Fenzi G, Grüters A, Busslinger M, Di Lauro R: PAX8 mutations associated with congenital hypothyroidism caused by thyroid dysgenesis. Nat Genet 1998;19:83–86.
11 Fujiwara H, Tatsumi K, Miki K, Harada T, Miyai K, Takai S, Amino N: Congenital hypothyroidism caused by a mutation in the Na^+/I^- symporter. Nat Genet 1997;16:124–125.
12 Everett LA, Glaser B, Beck JC, Idol JR, Buchs A, Heyman M, Adawi F, Hazani E, Nassir E, Baxevanis AD, Sheffield VC, Green ED: Pendred syndrome is caused by mutations in a putative sulphate transporter gene (PDS). Nat Genet 1997;17:411–422.
13 Scott DA, Wang R, Kreman TM, Sheffield VC, Karnishki LP: The Pendred syndrome gene encodes a chloride-iodide transport protein. Nat Genet 1999;21:440–443.
14 Stanbury JB, Rocmans P, Buhler UK, Ochi Y: Congenital hypothyroidism with impaired thyroid response to thyrotropin. N Engl J Med 1968;279:1132–1136.
15 Medeiros-Neto G, Herodotou DT, Rajan S, Kommareddi S, de Lacerda L, Sandrini R, Boguszewski MC, Hollenberg AN, Radovick S, Wondisford FE: A circulating, biologically inactive thyrotropin caused by a mutation in the beta subunit gene. J Clin Invest 1996;97: 1250–1256.
16 Gagné N, Parma J, Deal C, Vassart G, van Vliet G: Apparent congenital athyreosis contrasting with normal plasma thyroglobulin levels and associated with inactivating mutations in the thyrotropin receptor gene: Are athyreosis and ectopic thyroid distinct entities. J Clin Endocrinol Metab 1998;83:1771–7175.
17 Sunthornthepvarakui T, Gottschalk ME, Hayashi Y, Refetoff S: Brief report: Resistance to thyrotropin caused by mutations in the thyrotropin-receptor gene. N Engl J Med 1995;332: 155–160.
18 Clifton-Bligh RJ, Gregory JW, Ludgate M, John R, Persani L, Asteria C, Beck-Peccoz P, Chatterjee VK: Two novel mutations in the thyrotropin (TSH) receptor gene in a child with resistance to TSH. J Clin Endocrinol Metab 1997;82:1094–1100.

19 Kimura S, Hara Y, Pineau T, Fernandez-Salguero P, Fox CH, Ward JM, Gonzalez FJ: The T/ebp null mouse: Thyroid-specific enhancer-binding protein is essential for the organogenesis of the thyroid, lung, ventral forebrain, and pituitary. Genes Dev 1996;10:60–69.
20 De Felice M, Ovitt C, Biffali E, Rodriguez-Mallon A, Arra C, Anastassiadis K, Macchia PE, Mattei MG, Mariano A, Scholer H, Macchia V, Di Lauro R: A mouse model for hereditary thyroid dysgenesis and cleft palate. Nat Genet 1998;19:395–398.
21 Lapi P, Macchia PE, Chiovato L, Biffali E, Moschini L, Larizza D, Baserga M, Pinchera A, Fenzi G, Di Lauro R: Mutations in the gene encoding thyroid transcription factor-1 (TTF-1) are not a frequent cause of congenital hypothyroidism (CH) with thyroid dysgenesis. Thyroid 1997;7:383–387.
22 Devriendt K, Vanhole C, Matthijs G, de Zegher F: Deletion of thyroid transcription factor-1 gene in an infant with neonatal thyroid dysfunction and respiratory failure. N Engl J Med 1998;338:1317–1318.
23 Bamforth JS, Hughes IA, Lazarus JH, Weaver CM, Harper PS: Congenital hypothyroidism, spiky hair, and cleft palate. J Med Genet 1989; 26:49–51.
24 Buntincx IM, Van Overmeire B, Desager K, Van Hauwaert J: Syndromic association of cleft palate, bilateral choanal atresia, curly hair, and congenital hypothyroidism. J Med Genet 1993;30:427–428.
25 Lapi P, Macchia PE, Krude H, Chiovato L, Grüters A, Fenzi GF, Di Lauro R: Research for mutations in the TTF-2 gene in cases of congenital hypothyroidism due to thyroid dysgenesis. J Endocrinol Invest 1998;21:37.
26 Pfäffle RW, Parks JS, Brown MR, Ingraham HA, Di Mattia G, Wit JM: A mutation of pit-1 in two families with multiple pituitary hormone deficiencies. Eur J Pediatr 1992;151: 229–240.
27 Collu R, Tang J, Castagne J, Lagace G, Masson N, Huot C, Deal C, Delvin E, Faccenda E, Eidne KA, Van Vliet G: A novel mechanism for isolated central hypothyroidism: Inactivating mutations in the thyrotropin-releasing hormone receptor gene. J Clin Endocrinol Metab 1997;82:561–565.
28 Dacou-Voutetakis C, Feltquate DM, Drakopoulou M, Kourides IA, Dracopoli NC: Familial hypothyroidism caused by a nonsense mutation in the thyroid-stimulating hormone beta-subunit gene. Am J Hum Genet 1990;6:988–993.
29 Doeker BM, Pfäffle RW, Pohlenz J, Andler W: Congenital central hypothyroidism due to a homozygous mutation in the thyrotropin beta-subunit gene follows an autosomal recessive inheritance. J Clin Endocrinol Metab 1998;83: 1762–1765.
30 Grüters A, Liesenkötter KP, Zapico M, Jenner A, Dutting C, Pfeiffer E, Lehmkuhl U: Results of the screening program for congenital hypothyroidism in Berlin (1978–1995). Exp Clin Endocrinol Diabetes 1997;105(suppl 4):28–31.
31 Grüters A: Screening for congenital hypothyroidism: Effectiveness and clinical outcome. Baillières Clin Pediatr 1996;4:259–277.
32 Lazzaro D, Price M, de Felice M, Di Lauro R: The transcription factor TTF-1 is expressed at the onset of thyroid and lung morphogenesis and in restricted regions of the foetal brain. Development 1991;113:1093–1104.
33 Biebermann H, Liesenkötter KP, Emeis M, Oblanden M, Grüters A: Severe congenital hypothyroidism due to a homozygous mutation of the betaTSH gene. Pediatr Res 1999;46: 170–173.

Hormone Research

Horm Res 2000;53(suppl 1):19–25

Children with Septo-Optic Dysplasia – How to Improve and Sharpen the Diagnosis

Ann Hellström[a,b] Marita Aronsson[c] Cecilia Axelson[a] Mårten Kyllerman[c] Svenni Kopp[d] Susanne Steffenburg[d] Kerstin Strömland[b] Otto Westphal[a] Lars-Martin Wiklund[e] Kerstin Albertsson Wikland[a]

[a]Göteborg Paediatric Growth Research Centre, Department of Paediatrics; [b]Section of Ophthalmology, Department of Clinical Neuroscience; [c]Child Neurology, Department of Paediatrics, and Departments of [d]Neuropsychiatry and [e]Radiology, Queen Silvia Children's Hospital, Göteborg, Sweden

Key Words

Septo-optic dysplasia · Diagnosis · Children · Etiology, septo-optic dysplasia

Abstract

The data in this article are based on investigations performed in 25 children with suspected septo-optic dysplasia (SOD). There are many signs and methods that help in the diagnosis of SOD. In particular, the ocular fundus, abnormalities of the hypothalamo-pituitary axis and other midline brain structures should be described. In order to achieve a more holistic and functional diagnosis, the degree of neurological, neuropsychiatric and psychological involvement should also be stated. It has been suggested that SOD is associated with autosomal recessive inheritance, and it can be speculated that it is the result of genetic and environmental influences early in gestation. An early diagnosis can favourably influence the outcome of the affected child.

Introduction

Reeves [1] first described a septo-optic dysplasia syndrome in 1941. However, it was not until 15 years later that de Morsier [2] coined the expression septo-optic dysplasia (SOD), characterized by agenesis of the septum pellucidum and optic nerve hypoplasia (fig. 1). Another 15 years passed before Kaplan et al. [3] demonstrated that this syndrome is, to a high degree, associated with pituitary hormone insufficiencies.

In a number of reports, the frequencies of associated pituitary hormone insufficiencies have varied between 30 and 100% [4–11]. If these endocrinopathies remain untreated, the children might suffer from hypoglycaemia, adrenal crisis and even sudden death [12, 13]. This can be avoided by early correct diagnosis, although it was recently shown [14] that these children, as a group, are not treated with GH which is the first hormone to be affected, until they are over 5 years of age.

The diagnostic criteria of SOD have been inconsistent between different reports. As pointed out by Roessmann [6], the original concept of de Morsier [2] has been

KARGER

Fax + 41 61 306 12 34
E-Mail karger@karger.ch
www.karger.com

0301–0163/00/0537–0019$17.50/0

Accessible online at:
www.karger.com/journals/hre

Prof. K. Albertsson Wikland, GP-GRC
Department of Paediatrics, Queen Silvia Children's Hospital
S–416 85 Göteborg (Sweden)
Tel. +46 31 343 5167, Fax +46 31 848 952
E-Mail Kerstin.Albertsson-Wikland@pediat.gu.se

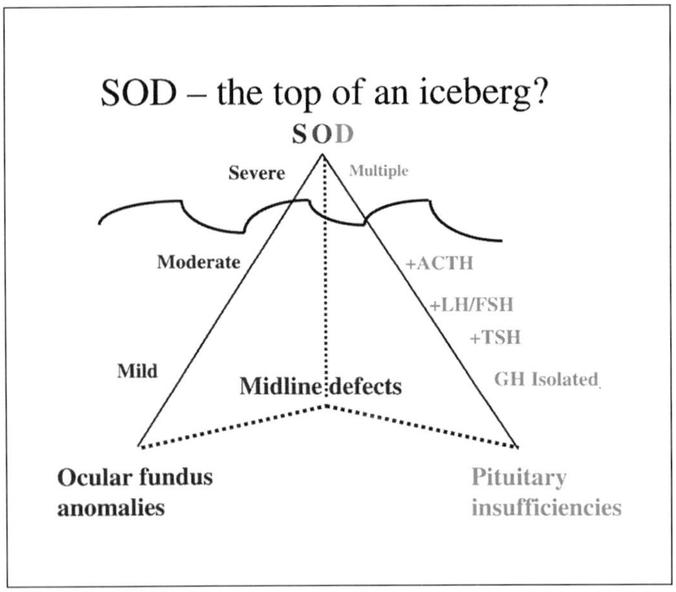

Fig. 1. Hypothetical illustration of the relationship between midline brain anomalies, optic nerve anomalies and pituitary anomalies.

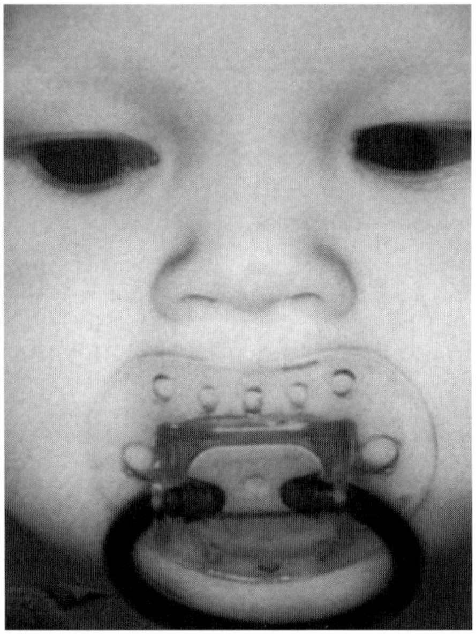

Fig. 2. A 2-year-old boy with SOD (absence of septum pellucidum, optic nerve hypoplasia and pituitary deficiencies) and hypertelorism.

Table 1. Neonatal and postnatal signs in children with SOD

Neonatal signs	Postnatal signs
Facial anomalies	Visual disturbances
Hypoglycaemia	Impaired growth
Hyperbilirubinaemia	Neurological
Micropenis	Neuropsychiatric

expanded and modified. In fact, the only criterion in common, in a number of studies on SOD patients, has been optic nerve hypoplasia [4]. Do these various midline abnormalities indeed constitute a spectrum of one and the same disorder?

In order to answer this question and map the aetiology of this heterogeneous syndrome, strict diagnostic criteria for each phenotype, including thorough description of the midline brain structures, the visual pathways and the pituitary region, are necessary.

Patients

During 1995–1999, the Göteborg Paediatric Growth Research Centre served as the national referral centre for children with suspected SOD through a Swedish Board of Health and Welfare programme. Twenty-five children were investigated, and 14 were diagnosed as having SOD (7 girls and 7 boys).

The data presented in this article are based upon investigations performed in this patient group as well as on a literature review.

Which Children Should Undergo Investigation?

Symptoms

Some important neonatal and postnatal signs in children with SOD are shown in table 1.

Neonatal Signs

The majority of children have a normal birth, although there is a slight increase of caesarian sections among children with multiple pituitary insufficiency (mostly due to breech delivery) as compared with children with isolated GH insufficiency [14]. The children, although born at

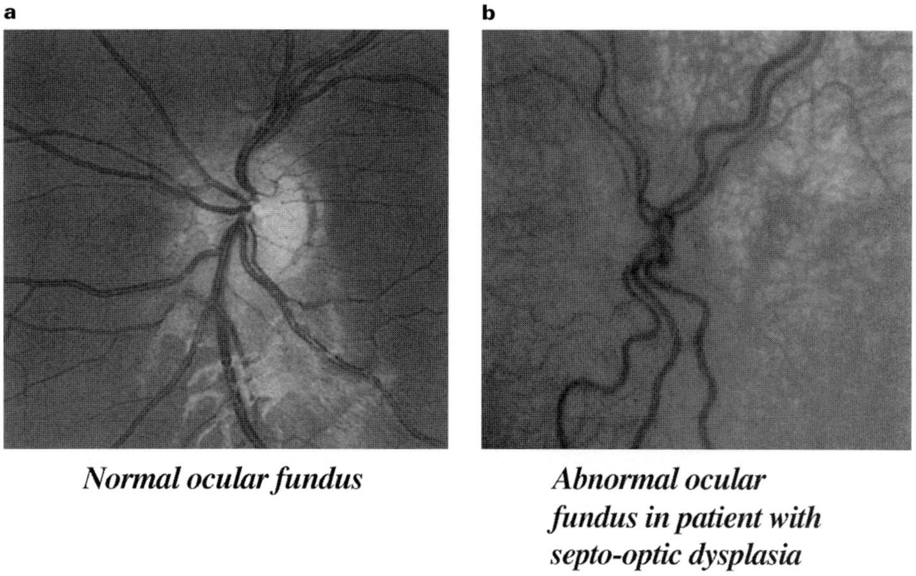

Fig. 3. **a** Normal ocular fundus. **b** Abnormal ocular fundus with optic nerve hypoplasia, small optic disc area, reduced retinal vascularization and isolated tortuosity of the retinal veins.

appropriate term and at full size, demonstrate, in a large proportion (20–60%), neonatal signs such as low Apgar scores, hypoglycaemia and hyperbilirubinaemia. Facial abnormalities (i.e., hypertelorism and underdeveloped midface structures) may be present, as is often noted among children with midline brain anomalies (fig. 2). Microgenitalism has been demonstrated in 28–50% of the boys [13, 14].

Postnatal Signs

Ophthalmological. Children with optic nerve hypoplasia and SOD have been reported to be blind in 25–50%, or severely visually impaired in another 40% [13–15]. A specific ocular fundus appearance with a markedly small optic disc and isolated tortuosity of the retinal veins was recently described by Hellström et al. [15] which might help the ophthalmologist identify patients with endocrinopathy in association with SOD (fig. 3). Optic disc dysplasia and morning glory sign have additionally been described in association with endocrinopathy, indicating that children with these ophthalmological abnormalities should undergo endocrinological investigation.

Postnatal Growth. During the first years of life, SOD children lose more than 2 SD in length (fig. 4). This has previously been observed among children with idiopathic congenital GH deficiency [16], a phenomenon most likely explained by the role of GH in infant growth.

Neurological. Seizures have been reported in 25–30% and psychomotor retardation in about 50–60% of these children [13, 17]. It should be noted that there is a broad

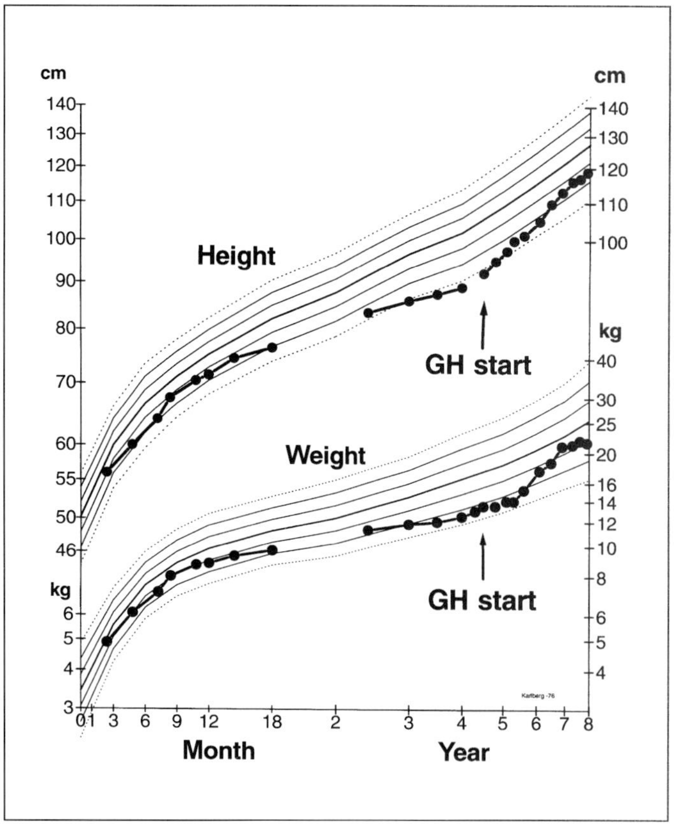

Fig. 4. Typical growth chart of an 8-year-old girl with SOD (absence of septum pellucidum, optic nerve hypoplasia and pituitary deficiencies).

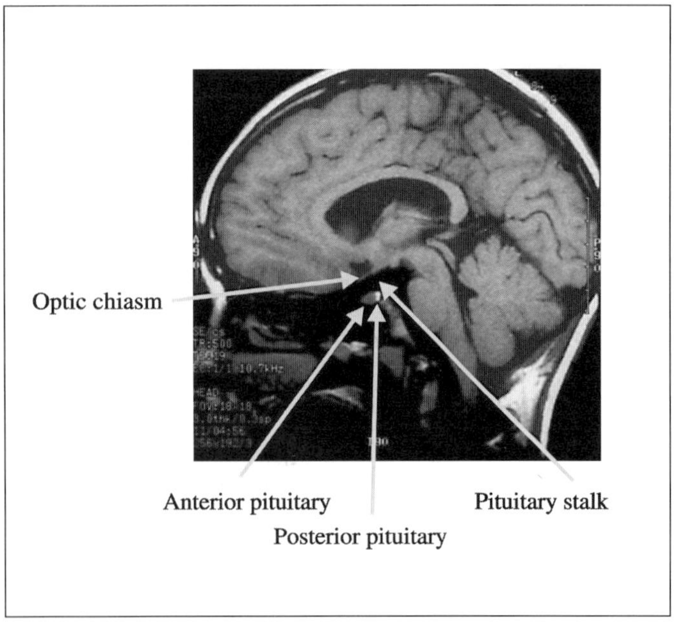

Fig. 5. Normal MRI of the pituitary region.

range of neurodevelopmental disabilities, ranging from normal intellect to severe mental retardation [13, 15].

Neuropsychiatric. Stereotypical behavioural abnormalities have been noted in a surprisingly high proportion (>50%) of our patients.

How Should a Child with SOD Be Investigated?

Examinations

Neuro-Imaging. The following questions need to be answered. (1) Is the anterior pituitary of normal size? (2) Is the posterior pituitary present, and what is its location? (3) Is the pituitary stalk present, and what is its size? (4) What are the sizes of optic nerves and optic chiasm? (5) Is the septum pellucidum present? (6) Are there other midline abnormalities (e.g. corpus callosum dysgenesis)?

Standardized MRI, focusing on the hypothalamus-pituitary region, including the visual pathways and septum pellucidum, should be performed in order to answer the above questions (fig. 5).

In our study, the following MRI setup has been used during neuro-imaging investigations: all MRI examinations were performed on a General Electric Signa 1.0-tesla magnet. Depending on the co-operabilty of the patient, the examination was performed with the patient either sedated or under general anesthesia. T1-weighted images were obtained in the sagittal and coronal planes, with and without gadolinium-DTPA enhancement. T2-weighted images were obtained in the coronal plane. The slice thickness was 3 mm with a scan interval of 0.3 mm. In addition, a routine T2-weighted scan of the entire brain was performed [18].

Ophthalmological. The following questions should be answered: (1) Does the child have optic nerve hypoplasia or some other optic disc anomaly? (Morphological signs to look for in optic nerve hypoplasia are: pallor of the optic disc, double-ring sign, small size of the optic disc and/or a small neuroretinal rim area. Other morphological abnormalities are: dysplastic optic discs or morning glory disc anomaly.) (2) Does the child have impaired vision? (Abnormal fixation behaviour or impaired visual acuity and/or visual fields should be assessed.) (3) Are there any motility disturbances (e.g., nystagmus or esotropia-unilateral optic nerve hypoplasia)?

Endocrinological. The following questions should be answered: (1) Is there normal somatotrophic function? (The GH response to provocation test or the spontaneous pulsatile GH secretion should be measured.) (2) Is there normal thyrotropic function? (Baseline serum thyroxine, free T_4 and T_3 levels and spontaneous thyroid-stimulating hormone or its response to thyrotropin-releasing hormone stimulation need to be assessed.) (3) Is there normal corticotropic function? (The spontaneous cortisol rhythm or stimulated serum cortisone should be checked.) (4) Is there normal gonadotropic function? (To measure are the basal luteinizing hormone and follicle-stimulating hormone levels and their stimulated levels in response to the luteinizing hormone releasing hormone test. Serum levels and rhythm of oestrogen and/or testosterone should also be assessed – these investigations depend on age.) (5) Is there normal lactotropic function? (Serum prolactin levels need to be measured.) (6) Is there normal posterior pituitary function? (All patients with polyuria and polydypsia should undergo a water dehydration test to confirm deficient vasopressin secretion, whereas the asymptomatic child with abnormal morphology of the posterior pituitary should be tested with early-morning paired serum and urine osmolality.)

Neurological and Neuropsychiatric. The following questions should be answered: (1) Does the child have seizures? (2) Does the child have neurodevelopmental disabilities? (3) Does the child have stereotypiced behavioural problems?

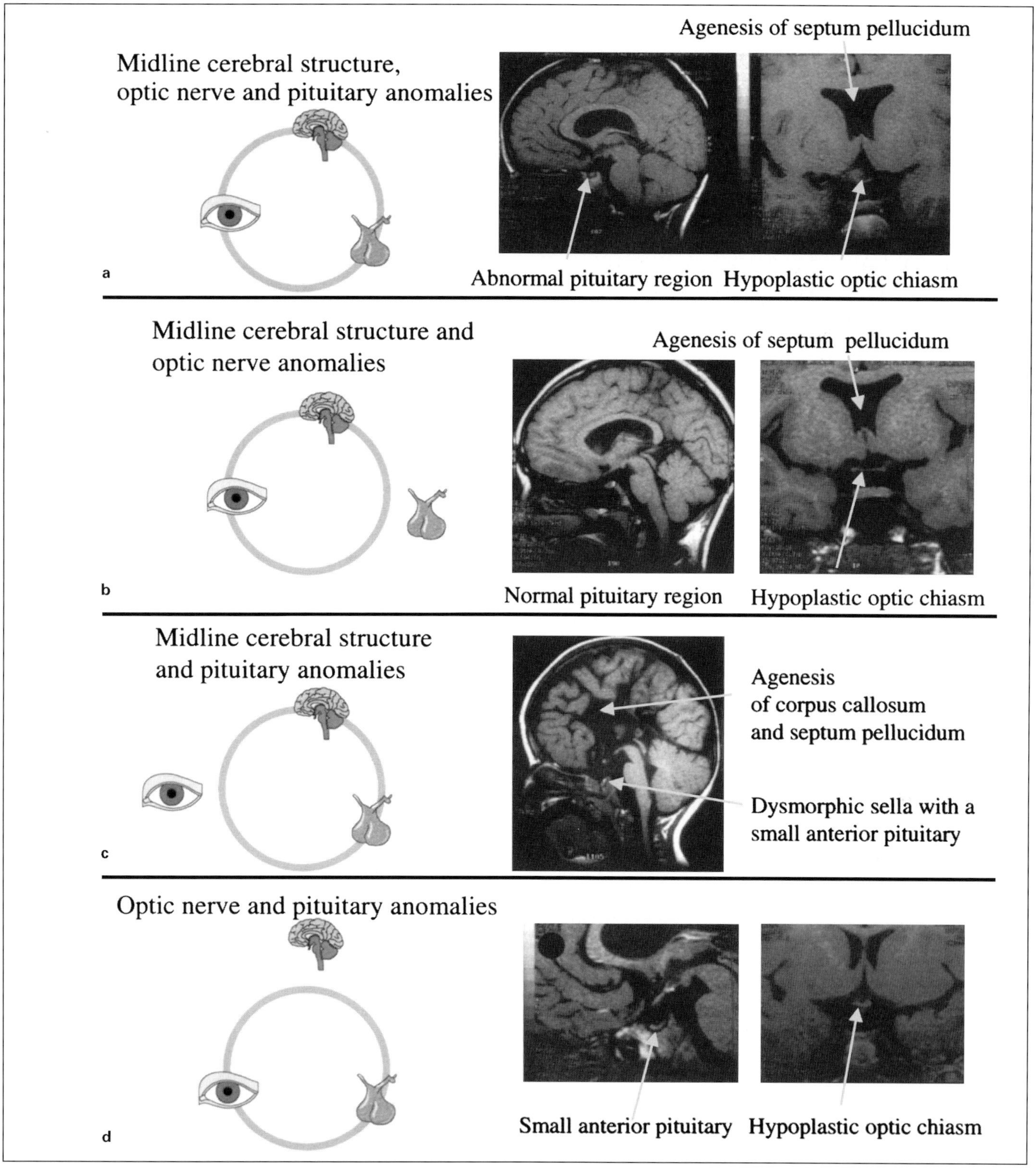

Fig. 6. a MRI of a 7-year-old blind boy with optic nerve hypoplasia, agenesis of the septum pellucidum and pituitary anomalies including GH deficiency. **b** MRI of a 13-year-old visually impaired girl with optic nerve hypoplasia and agenesis of the septum pellucidum. **c** MRI of a 5-year-old mentally retarded boy with dysgenesis of the corpus callosum and pituitary anomalies with GH deficiency. **d** MRI of a 6-year-old blind girl with optic nerve hypoplasia and pituitary anomalies with GH deficiency.

How Should a Child with SOD Be Diagnosed?

In order to have a proper and more specific diagnosis, a child should be investigated as suggested above. Thus, one should describe abnormalities of: (1) the hypothalamo-pituitary axis, including the extent of endocrinopathy; (2) the anterior visual pathways (optic nerve hypoplasia, coloboma or dysplasia), including the extent of visual impairment, and (3) other midline brain structures (e.g., septum pellucidum, corpus callosum).

In order to have a more holistic and functional diagnosis, the degree of neurological, neuropsychiatrical and psychological involvement should also be stated in each case.

Instead of using the term SOD, we would like to suggest the following anatomical diagnostic criteria: (1) midline cerebral structure, optic nerve and pituitary anomalies (fig. 6a); (2) midline cerebral structure and optic nerve anomalies (fig. 6b); (3) midline brain structures and pituitary anomalies (fig. 6c), and (4) optic nerve and pituitary anomalies (fig. 6d).

Aetiology

Embryologically, the diencephalon is the shared origin of optic nerve, hypothalamus and parts of the neurohypophysis. Maldevelopment of these structures has been believed to be due to an aberration during the early stages of development. At 4–6 weeks of gestation, the lamina terminalis thickens, and the optic and telencephalic vesicles begin to expand from the prosencephalon [19]. The retinal ganglion cells begin to form at 8 weeks of gestation (6 weeks after fertilization) [20], and maldevelopment of the septum pellucidum is thought to occur before 10 weeks of gestation [21]. A young maternal age has been a consistent finding in series of patients with SOD [22]. In addition, it has been questioned whether alcohol and drug toxicity may cause this syndrome [23]. A recent report has proposed that a vascular disruption, possibly involving the proximal trunk of the anterior cerebral artery, might be another explanatory factor for this syndrome [24]. Furthermore, an important role for the homeobox gene Hesx1/HESX1 in forebrain, midline and pituitary development in both mouse and human has been suggested to be associated with SOD [25].

It has been suggested that SOD may be associated with an autosomal recessive inheritance. Having the multifactorial aetiology in mind, one might speculate that the SOD syndrome reflects the result of genetic and environmental influences early in gestation. Thus, a vulnerable genotype might result in various phenotypes secondary to adverse environmental influences during a specific critical time period during development.

Acknowledgments

We are grateful for the support by the National Board of Health and Welfare/Sweden, to all paediatric endocrinologists who have provided us with information on SOD patients, for the support from the KIGS International Board, especially the strategic planning committee – Pierre Chatelain, Wayne Cutfield, David Anthony Price, Michael Ranke and Patrick Wilton –, and to the Medical Research Foundation for grants 7509, 10863 and 13515.

References

1 Reeves DL: Congenital absence of the septum pellucidum. Bull Johns Hopkins Hosp 1941; 69:61–71.
2 de Morsier G: Etudes sur les dystrophies cranioencéphaliques. III. Agénésie du septum lucidum avec malformation du tractus optique: la dysplasie septo-optique. Schweiz Arch Neurol Neurochir Psychiatr 1956;77:267–292.
3 Kaplan SL, Grumbach MM, Hoyt WF: A syndrome of hypopituitary dwarfism, hypoplasia of the optic nerves and malformation of the prosencephalon: Report of six patients. Pediatr Res 1970;4:480–481.
4 Willnow S, Kiess W, Butenandt O, Dorr HG, Enders A, Strasser-Vogel B, Egger J, Schwarz HP: Endocrine disorders in septo-optic dysplasia (De Morsier syndrome) – Evaluation and follow-up of 18 patients. Eur J Pediatr 1996; 155:179–184.
5 Stanhope R, Preece MA, Brook CGD: Hypoplastic optic nerves and pituitary dysfunction. Arch Dis Child 1984;59:111–114.
6 Roessmann U: Septo-optic dysplasia (SOD) or DeMorsier syndrome. J Clin Neuroophthalmol 1989;9:156–159.
7 Roessmann U, Velasco ME, Small EJ, Hori A: Neuropathology of 'septo-optic dysplasia' (de Morsier syndrome) with immunohistochemical studies of the hypothalamus and pituitary gland. J Neuropathol Exp Neurol 1987;46: 597–608.
8 Barkovich AJ, Fram EK, Norman D: Septo-optic dysplasia: MR imaging. Radiology 1989; 171:189–192.
9 Arslanian SA, Rothfus WE, Foley TP, Becker DJ: Hormonal, metabolic, and neuroradiologic abnormalities associated with septo-optic dysplasia. Acta Endocrinol (Copenh) 1984;107: 282–288.
10 Izenberg N, Rosenblum M, Parks JS: The endocrine spectrum of septo-optic dysplasia. Clin Pediatr (Phila) 1984;23:632–636.
11 Morishima A, Aranoff GS: Syndrome of septo-optic-pituitary dysplasia: The clinical spectrum. Brain Dev 1986;8:233–239.
12 Brodsky M, Conte FA, Taylor D, Hoyt CS, Mrak RE: Sudden death in septo-optic dysplasia. Arch Ophthalmol 1997;115:66–70.
13 Cameron FJ, Khadilkar VV, Stanhope R: Pituitary dysfunction, morbidity and mortality with congenital midline malformation of the cerebrum. Eur J Pediatr 1999;158:97–102.

14 Hellström A, Chatelain P, Cutfield W, Jönsson P, Price A, Wilton P, Albertsson Wikland K: Septo-optic dysplasia: Experience with 151 cases in KIGS (Pharmacia & Upjohn International Growth Database); in Ranke M, Wilton P (eds): Growth Hormone Therapy in KIGS: Ten Years' Experience 1987–1997. Heidelberg, JA Barth, 1999, vol 14, pp 147–157.
15 Hellström A, Wiklund L-M, Svensson E, Albertsson-Wikland K, Strömland K: Optic nerve hypoplasia with isolated tortuosity of the retinal veins – a marker of endocrinopathy. Arch Ophthalmol 1999;117:880–884.
16 Gluckman P, Gunn AJ, Wrai A, Cutfield WS, Chatelain PG, Guilbaud O, Ambler GR, Wilton P, Albertsson Wikland K on behalf of the International Board of the KABI Pharmacia International Growth Study: Congenital idiopathic growth hormone deficiency associated with prenatal and early postnatal growth failure. J Pediatr 1992;121:920–923.
17 Szhoof E, Walden U, Bergua A, Wenzel D, Dörr HG: Are endocrine disorders underestimated in patients with septo-optic dysplasia? 27th Int Symp on GH and Growth Factors in Endocrinology and Metabolism, Nice 1999.
18 Hellström A, Wiklund L-M, Svensson E, Strömland K, Albertsson-Wikland K: Midline brain lesions in children with hormone insufficiency indicate early prenatal damage. Acta Paediatr 1998;87:528–536.
19 O'Rahilly R, Müller F: The endocrine system; in O'Rahilly R, Müller F (eds): Human Embryology and Teratology, ed 2. New York, Wiley-Liss, 1996, pp 317–236.
20 Hendrickson AE: Primate foveal development: A microcosm of current questions in neurobiology. Invest Ophthalmol Vis Sci 1994;35:3129–3133.
21 Osborne AG: Handbook of Neuroradiology. St. Louis, Mosby, 1991, pp 167–174.
22 Elster AB, McAnarney ER: Maternal age and septo-optic dysplasia. J Pediatr 1979;94:162–163.
23 Coulter CL, Leech RW, Schaefer GB, Scheithauer BW, Brumback RA: Midline cerebral dysgenesis, dysfunction of the hypothalamic-pituitary axis, and fetal alcohol effects. Arch Neurol 1993;50:771–775.
24 Lubinsky MS: Hypothesis: Septo-optic dysplasia is a vascular disruption sequence. Am J Med Genet 1997;69:235–236.
25 Dattani MT, Martinez-Barbera JP, Thomas PQ, Brickman JM, Gupta R, Martensson IL, Toresson H, Fox M, Wales JK, Hindmarsh PC, Krauss S, Beddington RS, Robinson IC: Mutations in the homeobox gene HESX1/Hesx1 associated with septo-optic dysplasia in human and mouse. Nat Genet 1998;19:125–133.

Hormone Research

Horm Res 2000;53(suppl 1):26–33

Molecular Genetics of Septo-Optic Dysplasia

Mehul T. Dattani[a] Juan-Pedro Martinez-Barbera[b] Paul Q. Thomas[c] Joshua M. Brickman[b] Rajeev Gupta[b] Jeremy K.H. Wales[d] Peter C. Hindmarsh[a] Rosa S.P. Beddington[b] Iain C.A.F. Robinson[b]

[a]London Centre for Paediatric Endocrinology and Metabolism, Institute of Child Health, London, [b]National Institute for Medical Research, London, UK, [c]The Murdoch Institute, Melbourne, Australia, and [d]Sheffield Children's Hospital, Sheffield, UK

Key Words

Pituitary · Septo-optic dysplasia · *HESX1* · Homeobox · Rathke's pouch

Abstract

Septo-optic dysplasia (SOD) is a highly variable condition characterized by midline neurological abnormalities associated with pituitary hypoplasia and optic nerve hypoplasia. The aetiology is unknown. Mutant mice, in which a novel homeobox gene, *Hesx1*, has been disrupted, exhibit a phenotype that resembles the phenotype of SOD. We therefore wished to explore the possibility that this gene is implicated in SOD. We cloned and sequenced the human homologue *HESX1* and screened for mutations in affected individuals using single-stranded conformational polymorphism analysis, followed by cloning and sequencing of any exons which showed a band shift. Two siblings with SOD were homozygous for an Arg53Cys missense mutation within the HESX1 homeodomain, leading to a loss of in vitro DNA binding. Subsequently, we have identified heterozygous mutations in *HESX1* that are associated with milder pituitary phenotypes. Our studies indicate a vital role for *Hesx1*/*HESX1* in forebrain and pituitary development in mouse and man, and hence in some cases of SOD.

Septo-optic dysplasia (SOD)

SOD is a rare condition, which was initially described by Reeves [1] in a 7-month-old baby, and is often referred to as de Morsier syndrome [2]. It is a highly heterogeneous condition defined loosely by any combination of optic nerve hypoplasia, midline neuroradiological abnormalities, such as agenesis of the corpus callosum and absence of the septum pellucidum, and pituitary hypoplasia with consequent panhypopituitarism [1–7]. The phenotype is highly variable and may include midline defects without optic nerve hypoplasia. The diagnosis is usually made if two or more of the above features are present [8]. Pituitary hypoplasia may manifest as endocrine deficits varying from isolated growth hormone deficiency to panhypopituitarism. Either sexual precocity or failure to develop in puberty may occur, and abnormal hypothalamic neu-

KARGER

Fax + 41 61 306 12 34
E-Mail karger@karger.ch
www.karger.com

0301–0163/00/0537–0026$17.50/0

Accessible online at:
www.karger.com/journals/hre

Dr. M.T. Dattani
London Centre for Paediatric Endocrinology
Institute of Child Health, 30 Guilford Street
London WC1N 1EH (UK)
Tel. +44 171 242 9789, Fax +44 171 404 6191, E-Mail M.Dattani@ich.ucl.ac.uk

roanatomy or function may be a feature [9, 10], as may diabetes insipidus [11]. The endocrinopathy may be evolving with a progressive loss of endocrine function over time [8]. The commonest endocrinopathies are growth hormone deficiency and adrenocorticotrophic hormone (ACTH) deficiency [12, 13]. Gonadotrophin secretion may be retained in the face of other pituitary hormone deficiencies [14], and these findings suggest a complex aetiology with hypothalamic and pituitary components in many of these patients.

Several aetiologies have been postulated to account for the sporadic occurrence of SOD, such as viral infections, environmental teratogens, and vascular or degenerative damage [10, 15]. Familial cases of SOD are rare and may be associated with an autosomal recessive inheritance [16–18]. We have recently identified a murine model for SOD [19]. The phenotype of this mouse was the result of targeted disruption of a novel murine homeobox gene, *Hesx1*. Homeobox genes play a crucial role in regional patterning and cell differentiation during the development of multicellular organisms. They were first described in *Drosophila* and are characterized by the presence of a homeobox that encodes a highly conserved 60 amino acid homeodomain at the protein level. Protein products of these genes can, therefore, act as specific transcription factors which bind to regulatory elements of target genes by means of the homeodomain. The homeobox gene family can be subdivided into about 30 classes of genes, each characterized by its own specific consensus sequence of the homeodomain [20]. Mutations in several homeobox genes have been identified as the molecular defects in several human anomalies, including *PAX2* (renal coloboma syndrome) [21], *PAX3* (Waardenberg syndrome type I) [22–24], *PAX6* (aniridia) [25], *MSX1* (selective tooth agenesis) [26], *MSX2* (craniosynostosis) [27], *HOXD13* (limb deformities) [28], *PIT1* and *PROP1* (hypopituitarism and combined hormone deficiencies) [29–31] and *SHOX* (short stature) [32]. The role of homeobox genes in development is elegantly demonstrated by the pituitary gland (see below).

Development of the Anterior Pituitary Gland

The pituitary gland, which is a midline structure, consists of the anterior, intermediate and posterior lobes and has a dual embryonic origin. The origins can be traced back to the anterior neural ridge and the neural plate [33, 34]. The anterior and intermediate lobes derive from Rathke's pouch which is itself an invagination of the midline oral ectoderm and lies in direct contact with the diencephalon. The posterior lobe is derived as a result of the evagination of the neural ectoderm (infundibulum) at the base of the developing diencephalon in direct contact with Rathke's pouch. Hence, in the mouse, the anterior and intermediate lobes of the pituitary gland form on embryonic day 8.5 (E8.5) from Rathke's pouch, and this makes direct cell-cell contact with the neuro-epithelium of the nascent diencephalon at E8.5–9.0 [35]. The apposition of Rathke's pouch and the diencephalon is maintained throughout the early stages of pituitary organogenesis, and this close relationship together with various animal experimental manipulations has long suggested that inductive tissue interactions are involved in the process. The development of the anterior pituitary gland is dependent on various intrinsic and extrinsic factors. The latter include *Ttf1*, *BMP4* and *Fgf8* [36–38] which are expressed in the diencephalon and not in Rathke's pouch, and a role for which has been suggested in normal pituitary development.

It has now been established that Rathke's pouch develops in a two-step process which requires at least two sequential inductive signals from the diencephalon. First, induction and formation of the pouch rudiment is dependent upon BMP4 which is present only in the hypothalamus and not in Rathke's pouch. Second, FGF8, which is also present in the hypothalamus and not in Rathke's pouch, activates the key regulatory genes *Lhx3* and *Lhx4* (see below), and these are essential for subsequent development of the pouch rudiment into a definitive pouch [38].

A number of transcription factors are expressed within the pituitary primordium. These include *Six3* [39], *Pax6* [40] and *Hesx1/Rpx* [41, 42], all of which continue to be expressed in Rathke's pouch. Concurrent with organ commitment, two LIM homeodomain factors, *Lhx3* or *P-Lim* [43, 44] and *Lhx4* [45], are also expressed in Rathke's pouch. Deletion of the *Lhx3/P-Lim* gene in mice results in the failure of the pituitary gland to grow and differentiate, even though formation of Rathke's pouch does occur [46]. However, a double *Lhx3/Lhx4* mutant results in the failure of formation of Rathke's pouch [45]. The OTX-related factor *Ptx1/P-OTX* is also expressed throughout early pituitary development [47, 48].

The mature anterior pituitary gland is populated by five neuroendocrine cell types, each defined by the hormone produced: corticotropes (ACTH), thyrotropes (thyroid-stimulating hormone, TSH), gonadotropes (luteinizing hormone, LH, follicle-stimulating hormone, FSH), somatotropes (growth hormone, GH) and lactotropes (prolactin, PRL). Each of these cell types differentiates in a temporally and spatially regulated manner [49, 50], and this process is dependent upon a number of transcription factors. A nuclear receptor, SF1, is expressed later in the nascent gonadotrope lineage [51]. A tissue-specific POU domain factor, Pit1, is required for terminal differentiation, growth and survival of somatotropes, lactotropes and thyrotropes [52]. Additionally, a paired-like homeodomain factor, Prop1, is also required for the determination of the Pit1-dependent lineage [53]. Mutations of both *Pit1* and *Prop1* have been described in man and account for the phenotype of GH, PRL and TSH deficiencies [29, 30] and GH, PRL, TSH and FSH/LH deficiencies [31], respectively. These findings suggest that Prop1 is essential for determination of the gonadotrope lineage as well as the Pit1-dependent lineage in man.

The Homeobox Gene *Hesx1*

Hesx1 is a member of the paired-like class of homeobox genes [41] and is first expressed during mouse embryogenesis in a small patch of cells in the anterior midline visceral endoderm, as gastrulation commences [42, 54]. Expression is then induced in the adjacent ectoderm in an area destined to give rise to the ventral prosencephalon which will, in turn, form the forebrain [55]. This ectodermal expression appears to require an inductive signal from the visceral endoderm [54]. Subsequently, *Hesx1* is expressed at the anterior extreme of the rostral neural folds; expression is then restricted to the ventral diencephalon by E9.0. At this stage, *Hesx1* is also expressed in the thickened layer of the oral ectoderm which will give rise to Rathke's pouch. *Hesx1* continues to be expressed in the developing anterior pituitary until E12, when its transcripts disappear in a spatiotemporal sequence that corresponds to progressive pituitary cell differentiation. Expression is undetectable by E13.5.

In the Ames dwarf mouse, the phenotype of which arises as a result of a mutation in the *Prop1* gene, *Hesx1* expression persists until E16, suggesting a role for *Prop1* in regulating *Hesx1* expression [53]. Additionally, a role for *Lhx3* in the maintenance of *Hesx1* expression has been suggested by the *Lhx3* null mutants [46]. As *Hesx1* is a putative repressor gene [19], its role in pituitary development may be to repress downstream target genes. No targets of *Hesx1* have been identified to date.

Targeted disruption of *Hesx1* in the mouse revealed morphological defects in homozygote embryos from E9 and, also, in mutant neonates and adults [19]. These included a reduction in prospective forebrain tissue, absence of developing optic vesicles, markedly decreased head size, craniofacial dysplasia with a short nose and severe microphthalmia. The defects were often asymmetrical, with one side more severely affected than the other. By E10, the severity of the mutant phenotype was variable, and mutants could be divided into either a class I phenotype, which was severe, or a milder class II phenotype. All homozygotes were abnormal at birth, as were 1% of the heterozygotes. Histological analysis of the mutant mice revealed that class I embryos at E13 had substantially reduced forebrains with no sign of telencephalic vesicle or infundibulum development. Additionally, absence of the optic cups, the olfactory placodes and Rathke's pouch was noted. Class II mutant embryos demonstrated a thin forebrain epithelium with retardation in prosencephalic development. Other abnormalities included reduced telencephalic vesicles with hypothalamic abnormalities and aberrant morphogenesis of Rathke's pouch. The olfactory pits and optic vesicles were also hypoplastic and often asymmetric in appearance.

Further analysis of neonatal and adult mutants revealed hypoplastic nasal cavities, hypoplastic olfactory bulbs, microphthalmia and anophthalmia. Abnormalities of septum pellucidum and corpus callosum were also noted (fig. 1). The variable forebrain midline defects associated with abnormal morphogenesis of the anterior pituitary gland in the mutant mice were reminiscent of the syndrome of SOD in man [19].

In the light of the phenotype demonstrated in *Hesx1* null mutant mice, we investigated the role of the human homologue of this gene in patients with SOD. We isolated the human homologue *HESX1* from a human genomic library. The gene was cloned and sequenced and mapped to chromosome 3p21.1–3p21.2. The *HESX1* locus spans 1.7 kb with a highly conserved genomic organization consisting of four coding exons. The 185 amino acid open reading frame (ORF) is highly conserved as compared with the mouse and *Xenopus* ORFs, particularly across the homeodomains that share 95% and 80% homology, respectively. An additional conserved domain is located near the amino terminus which is also present in other homeodomain classes including *engrailed* and *goosecoid* and has been postulated to play a role in transcriptional repression.

We have now screened over 150 patients with a variety of pituitary disorders. Southern blot analysis performed in 38 of these patients revealed no deletions or gross rearrangements associated with the phenotype. Our patient population included a highly consanguineous family in

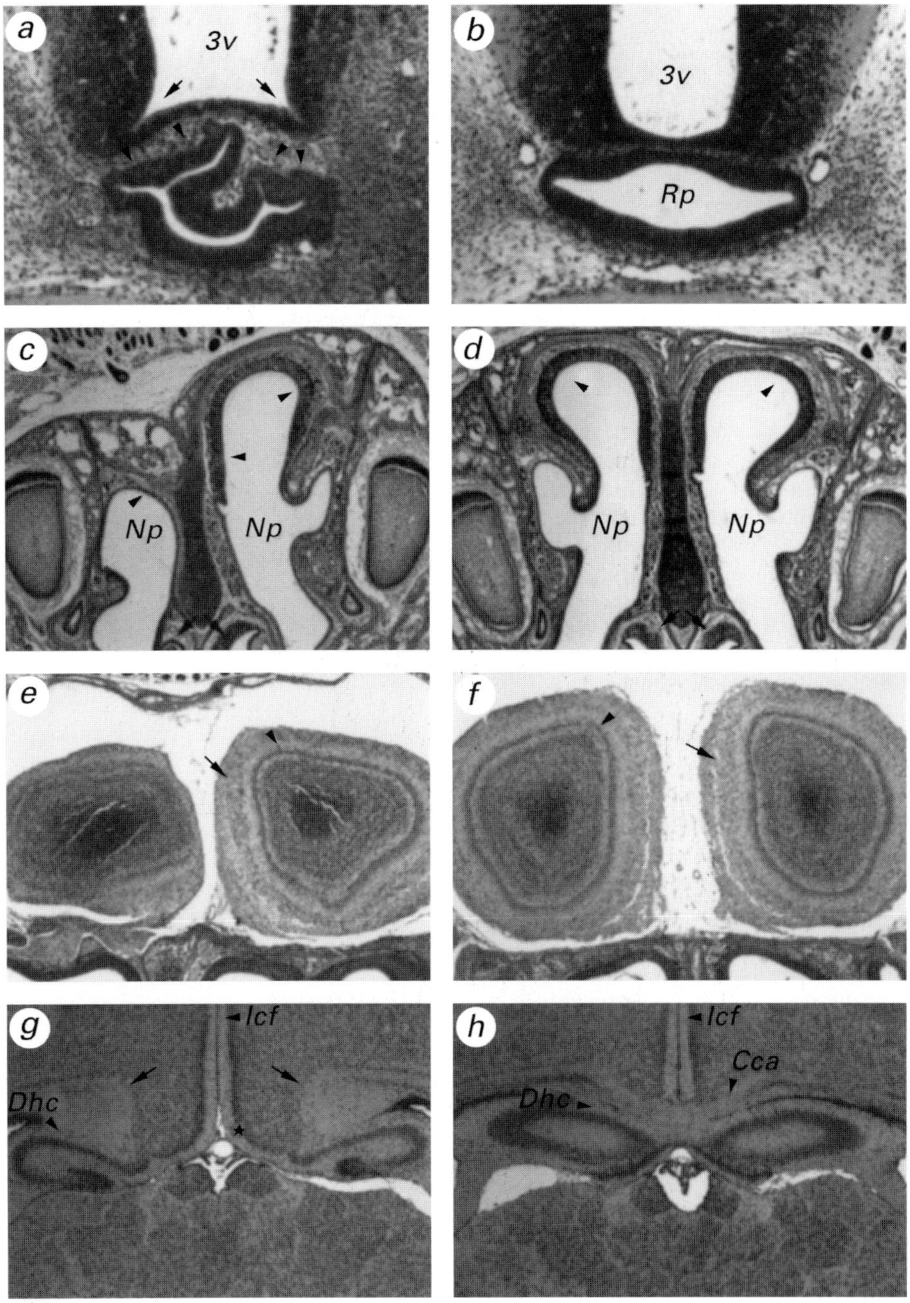

Fig. 1. Histological analysis of the *Hesx1* mutants. Transverse sections through hypothalamus and Rathke's pouch of 12.5-day post coitum mutant (**a**) and wild-type (**b**) embryos. Lateral expansion of Rathke's pouch (arrowheads in **a**) and bifurcation of the third ventricle (arrows in **a**) are evident in the mutant embryo but not in the wild-type littermate. **c–h** Transverse sections through the head of mutant (**c**, **e**, **g**) and wild-type (**d**, **f**, **h**) newborn mice at the levels of the nasal cavities (**c**, **d**), olfactory bulbs (**e**, **f**) and anterior hippocampus (**g**, **h**). **c**, **d** The left nasal passage in the mutant is atrophic and no nasal neuroepithelium (arrowheads) or vomerulonasal organ (arrows) has developed on this side. The right nasal passage appears normal as compared with the wild type. **e**, **f** The left olfactory bulb is hypoplastic relative to the wild type, and no glomerular (arrows) or mitral (arrowheads) cell layers can be observed on this side. Note the asymmetrical development of both the nasal passages and the olfactory bulbs in the *Hesx1* mutant as compared with the wild-type littermate. **g**, **h** Sections through corpus callosum and dorsal hippocampal commissure of a mutant (**g**) and wild-type (**h**) brain. **g** 8-day-old mutant neonate showing absence of the corpus callosum, though nerve fibres have formed Probst bundles on either side of the intercerebral fissure (arrows). The dorsal hippocampal commissure also does not cross the midline in the mutant and the intercerebral fissure appears continuous with the third ventricle (asterisk). Cca = Corpus callosum; Dhc = dorsal hippocampal commissure; Icf = intercerebral fissure; Np = nasal passage; Rp = Rathke's pouch; 3v = third ventricle. Reproduced with permission from Dattani et al. [19].

which 2 affected siblings presented with optic nerve hypoplasia, absence of the corpus callosum and hypoplasia of the pituitary gland with panhypopituitarism [16]. Single-stranded conformational polymorphism analysis using primers flanking each of the four *HESX1*-coding exons was used to screen these patients for point mutations within the *HESX1* ORF. We found a homozygous missense mutation in the homeobox of *HESX1* in the 2 affected siblings which resulted in the substitution of a highly conserved arginine residue (Arg53) by cysteine. The parents were heterozygous for the mutation. This mutation led to the loss of a *Cac*8I restriction site (fig. 2a), and hence, restriction analysis of a polymerase chain reaction product spanning exons 3 and 4 was used to screen for the mutated allele within the extended family (fig. 2b). This confirmed that the affected siblings IV.4 and IV.5

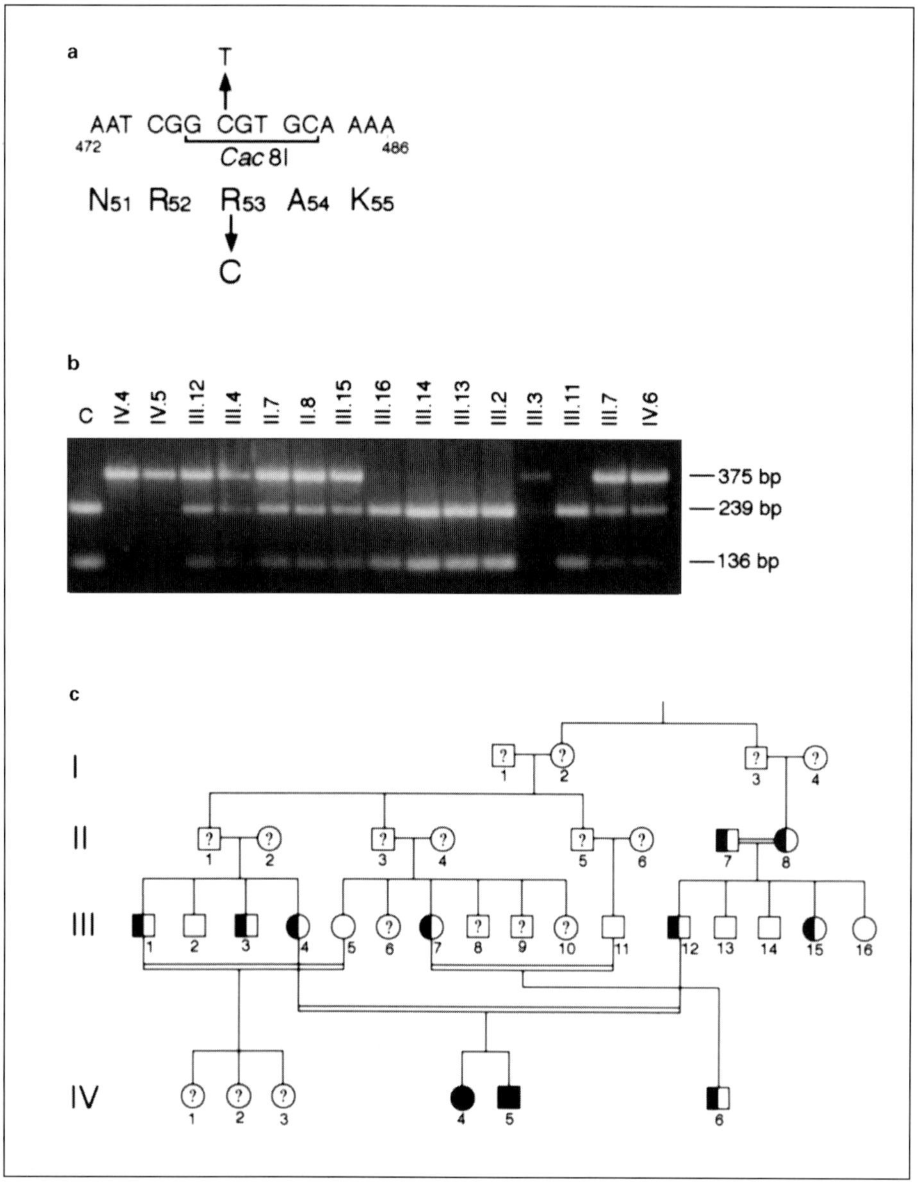

Fig. 2. Detection of a homozygous missense mutation in *HESX1* of the probands IV.4 and IV.5. **a** A C478T substitution in exon 4 destroys the *Cac81* restriction site and results in an Arg53Cys substitution within the *HESX1* homeodomain. **b** Genotype analysis using *Cac81* digests of PCR products from the extended family. Normal individuals give two bands (262 and 126 bp). Individuals homozygous for the mutation have a single (undigested) 388-bp band, while heterozygotes show three bands (388, 262 and 126 bp) The control sample for an unaffected unrelated individual is shown in the first lane (C). **c** Family pedigree of the probands IV.4 and IV.5. The filled symbols represent individuals homozygous for the mutation, the half-filled symbols represent heterozygotes and open symbols represent individuals with a homozygous wild-type genotype. The question mark refers to individuals not tested. III.1 and III.5 were genotyped in a separate *Cac81* assay (not shown in **b**). Note the multiple consanguinous unions (indicated with double lines) in this kindred. Reproduced with permission from Dattani et al. [19].

were homozygous for the mutation and revealed nine phenotypically normal heterozygotes within this highly consanguineous pedigree, consistent with an autosomal recessive inheritance (fig. 2c). Loss of the *Cac*8I restriction site was not observed in 10 unrelated patients with sporadic SOD or pituitary hypoplasia, nor in 70 normal individuals.

The Arg53 residue is one of the most highly conserved residues in homeodomain proteins from a broad range of eukaryotic species. It has been shown to contact the DNA phosphate backbone in the paired and engrailed DNA crystal structures [56, 57]. A non-conservative substitution of this amino acid by cysteine would be expected to compromise DNA binding. To test this, recombinant wild-type and mutant *HESX1* proteins were purified (fig. 3a) and their DNA binding activity assayed in an electromobility shift assay, using the palindromic PIII sequence to which murine *Hesx1* has previously been shown to bind [53]. Wild-type *HESX1* binds this sequence as a dimer with a high-affinity even at concentrations of 1 n*M*, whereas the mutant protein fails to bind even at micromolar concentrations (fig. 3b). The severe reduction in in vitro PIII binding activity of the mutant HESX1 protein compared with the wild type protein sug-

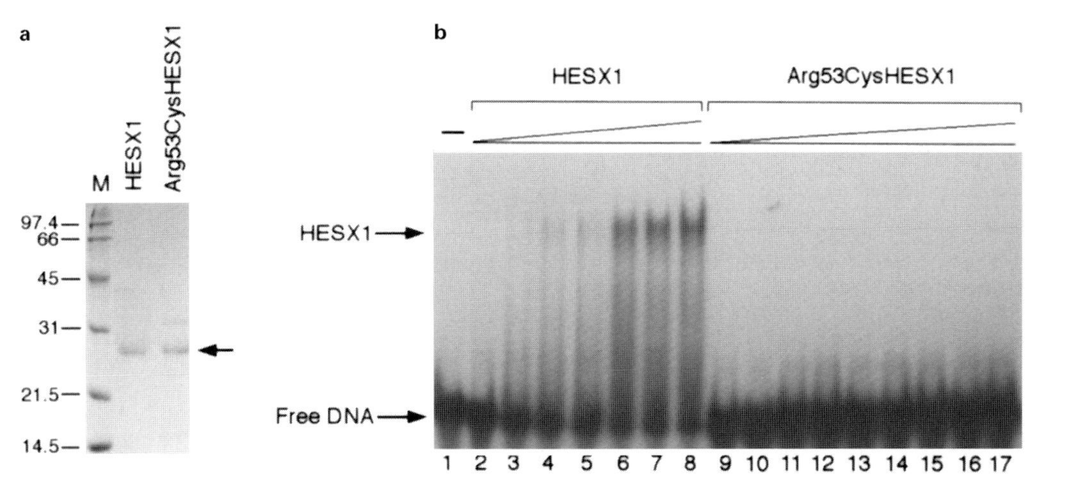

Fig. 3. Electromobility shift assay of wild-type and mutant *HESX1*. **a** Coomassie blue stain of wild-type (*HESX1*) and mutant (*Arg53CysHESX1*) recombinant proteins. Full-length *HESX1* protein is approximately 28 kD. **b** DNA binding by wild-type and *Arg53Cys HESX1* to an oligonucleotide containing a palindromic PIII site (5′-AGCTTGAGTC-TAATTGAATTAACTGTAC-3′). Lane 1 contains only DNA probe. Lanes 2–8 contain increasing amounts of wild-type recombinant *HESX1* (0.9, 1.8, 3.6, 7.1, 14.2, 28 and 57 n*M*). Lanes 9–17 contain increasing amounts of Arg53Cys *HESX1* protein (1.1, 2.3, 4.5, 9.1, 18, 36, 72, 144 and 288 n*M*). Reproduced with permission from Dattani et al. [19].

gests that the Arg53Cys substitution may act as a null mutation in vivo.

We have subsequently found five additional heterozygous mutations of *HESX1* in children with various forms of pituitary disease and SOD. Preliminary data suggest that at least two of these mutations have functional significance. Additionally, the inheritance appears to be autosomal dominant with a variable penetrance, as was observed in a small proportion of the heterozygous mutant mice [manuscript in preparation]. The situation with these mutations may be analogous to the R271W dominant negative mutation in the *PIT1* gene in a family with GH, PRL and TSH deficiencies [58].

Conclusions

Mutations in homeobox genes have been shown to play a role in the pathogenesis of a number of conditions in man. We have demonstrated a role for the homeobox gene *Hesx1* in a rare familial form of SOD. Our data also suggest a role for this gene in some cases of the more frequently occurring sporadic form of SOD. Alterations in regulatory regions may also contribute to the aetiology of this condition. Additionally, SOD may have a multigenic basis, to which alterations in *HESX1* could contribute. To date, the targets and partners of *Hesx1/HESX1* remain unknown. These may shed further light on the molecular basis not only of SOD, but of various other pituitary disorders. Whatever the mechanism, the striking resemblance between the mouse and human phenotype suggests that the gene may play a similar role in human and mouse embryonic development.

References

1 Reeves DL: Congenital absence of septum pellucidum. Bull Johns Hopkins Hosp 1941;69: 61–71.
2 de Morsier G: Etudes sur les dysraphies cranio-encéphaliques. III. Agénésie du septum lucidum avec malformation du tractus optique: la dysplasie septo-optique. Schweiz Arch Neurol Psychiatr 1956;77:267–292.
3 Hoyt WF, Kaplan SL, Grumbach MM, Glaser JS: Septo-optic dysplasia and pituitary dwarfism. Lancet 1970;i:893–894.
4 Brook CGD, Sanders MD, Hoare RD: Septo-optic dysplasia. Br Med J 1972;iii:811–813.
5 Arslanian SA: Hormonal, metabolic and neuroradiologic abnormalities associated with septo-optic dysplasia. Acta Endocrinol (Copenh) 1984;107:282–288.
6 Izenberg N, Rosenblum M, Parks JS: The endocrine spectrum of septo-optic dysplasia. Clin Pediatr (Phila) 1984;23:632–636.
7 Roessman U: Septo-optic dysplasia (SOD) or de Morsier syndrome. J Clin Neuroophthalmol 1989;9:156–159.
8 Stanhope R, Preece MA, Brook CGD: Hypoplastic optic nerves and pituitary dysfunction: A spectrum of anatomical and endocrine abnormalities. Arch Dis Child 1984;59:111–114.
9 Lam KSL, Wang C, Ma JT, Leung SP, Yeung RT: Hypothalamic defects in two adult patients with septo-optic dysplasia. Acta Endocrinol (Copenh) 1986;112:305–309.
10 Roesmann U, Velasco ME, Small EJ, Hori A: Neuropathology of 'septo-optic dysplasia' (de Morsier syndrome) with immunohistochemical studies of the hypothalamus and pituitary gland. J Neuropathol Exp Neurol 1987;64: 597–608.
11 Masera N, Grant DB, Stanhope RG, Preece MA: Diabetes insipidus with impaired osmotic regulation in septo-optic dysplasia and agenesis of the corpus callosum. Arch Dis Child 1994; 70:51–53.
12 Willnow S, Kiess W, Butenandt O, Dorr HG, Enders A, Strasser-Vogel B, Egger J, Schwarz HP: Endocrine disorders in septo-optic dysplasia (de Morsier syndrome) – evaluation and follow up of 18 patients. Eur J Pediatr 1996;155: 179–184.
13 Cameron FJ, Khadilkar V, Stanhope R: Pituitary dysfunction, morbidity and mortality with congenital midline malformation of the cerebrum. Eur J Pediatr 1999;158:97–102.
14 Nanduri VR, Stanhope R: Why is the retention of gonadotrophin secretion common in children with panhypopituitarism due to septo-optic dysplasia? Eur J Endocrinol 1999;140: 48–50.
15 Zaias B, Becker D: Septo-optic dysplasia: Developmental or acquired abnormality? A case report. Trans Am Neurol Assoc 1978;103:273–277.
16 Wales JKH, Quarrell OWJ: Evidence for possible mendelian inheritance of septo-optic dysplasia. Acta Paediatr 1996;85:391–392.
17 Blethen SL, Weldon VV: Hypopituitarism and septo-optic dysplasia in first cousins. Am J Med Genet 1985;21:123–129.
18 Benner JD, Preslan MW, Gratz E, Joslyn J, Schwartz M, Kelman S: Septo-optic dysplasia in two siblings. Am J Ophthalmol 1990;109: 632–637.
19 Dattani MT, Martinez-Barbera JP, Thomas PQ, Brickman JM, Gupta R, Martensson I, Toresson H, Fox M, Wales JKH, Hindmarsh PC, Krauss S, Beddington RSP, Robinson ICAF: Mutations in the homeobox gene *HESX1/Hesx1* associated with septo-optic dysplasia in human and mouse. Nat Genet 1998; 19:125–133.
20 Kappen C, Schughart K, Ruddle FH: Early evolutionary origin of major homeodomain sequence classes. Genomics 1993;18:54–70.
21 Sanyanusin P, Schimmenti LA, McNoe LA, Ward TA, Pierpont MEM, Sullivan MJ, Dobyns WB, Eccles MR: Mutation of the *PAX2* gene in a family with optic nerve colobomas, renal anomalies and vesicoureteral reflux. Nat Genet 1995;9:358–363.
22 Baldwin CT, Hoth CF, Amos JA, da Silva EO, Milunsky A: An exonic mutation in the *HuP2* paired domain gene causes Waardenburg's syndrome [see comments]. Nature 1992;355:637–638.
23 Morell R, Friedman TB, Asher JH Jr: A plus-one frameshift mutation in *PAX3* alters the entire deduced amino acid sequence of the paired box in a Waardenburg syndrome type 1 (WS1) family. Hum Mol Genet 1993;2:1487–1488.
24 Tassabehji M, Read AP, Newton VE, Harris R, Balling R, Gruss P, Strachan T: Waardenburg's syndrome patients have mutations in the human homologue of the *Pax-3* paired box gene [see comments]. Nature 1992;355:635–636.
25 Glaser T, Walton DS, Maas RL: Genomic structure, evolutionary conservation and aniridia mutations in the human *PAX6* gene. Nat Genet 1992;2:232–239.
26 Vastardis H, Karimbux N, Guthua SW, Seidman JG, Seidman CE: A human *MSX1* homeodomain missense mutation causes selective tooth agenesis [see comments]. Nat Genet 1996;13:417–421.
27 Jabs EW, Muller U, Li X, Ma L, Luo W, Haworth IS, Klisak I, Sparkes R, Warman ML, Mulliken JB, Snead ML, Maxson R: A mutation in the homeodomain of the human *MSX2* gene in a family affected with autosomal dominant craniosynostosis. Cell 1993;75:443–450.
28 Muragaki Y, Mundlos S, Upton J, Olsen BR: Altered growth and branching patterns in synpolydactyly caused by mutations in *HOXD13*. Science 1996;272:548–551.
29 Pfaffle RW, Di Mattia GE, Parks JS, Brown MR, Wit JM, Jansen M, Van der Nat H, Van den Brande JL, Rosenfeld MG, Ingraham HA: Mutation of the POU-specific domain of Pit-1 and hypopituitarism without pituitary hypoplasia. Science 1992;257:1118–1121.
30 Radovick S, Nations M, Du Y, Berg LA, Weintraub BD, Wondisford FE: A mutation in the POU-Homeodomain of Pit-1 responsible for combined pituitary hormone deficiency. Science 1992;257:1115–1118.
31 Wu W, Cogan JD, Pfäffle RW, Dasen JS, Frisch H, O'Connell MG, Flynn SE, Brown MR, Mullis PE, Parks JS, Phillips JA III, Rosenfeld MG: Mutations in *PROP1* cause familial combined pituitary hormone deficiency. Nat Genet 1998;18:147–149.
32 Rao E, Weiss B, Fukami M, Rump A, Niesler B, Mertz A, Muroya K, Binder G, Kirsch S, Winkelmann M, Nordsiek G, Heinrich U, Breuning MH, Ranke MB, Rosenthal A, Ogata T, Rappold GA: Pseudoautosomal deletions encompassing a novel homeobox gene cause growth failure in idiopathic short stature and Turner syndrome. Nat Genet 1997;16:54–63.
33 Couly G, Le Douarin NM: The fate map of the cephalic neural primordium at the presomitic to the 3-somite stage in the avian embryo. Development 1988;103(suppl):101–113.
34 Eagleson GW, Harris WA: Mapping of the presumptive brain regions in the neural plate of *Xenopus laevis*. J Neurobiol 1990;21:427–440.
35 Jacobsen AG, Miyamoto DM, Mai SH: Rathke's pouch morphogenesis in the chick embryo. J Exp Zool 1979;207:351–366.
36 Kimura S, Hara Y, Pineau T, Fernandez-Salguero P, Fox CH, Ward JM, Gonzalez FJ: The T/ebp null mouse: Thyroid-specific enhancer-binding protein is essential for the organogenesis of the thyroid, lung, ventral forebrain, and pituitary. Genes Dev 1996;10:60–69.
37 Treier M, Gleiberman AS, O'Connell SM, Szeto DP, McMahon JA, McMahon AP, Rosenfeld MG: Multistep signaling requirements for pituitary organogenesis in vivo. Genes Dev 1998;12:1691–1704.
38 Takuma N, Sheng HZ, Furuta Y, Ward JM, Sharma K, Hogan BLM, Pfaff SL, Westphal H, Kimura S, Mahon KA: Formation of Rathke's pouch requires dual induction from the diencephalon. Development 1998;125:4835–4840.
39 Oliver G, Wehr R, Jenkins NA, Copeland NG, Cheyette BN, Hartenstein V, Zipursky SL, Gruss P: Homeobox genes and connective tissue patterning. Development 1995;121:693–705.
40 Walther C, Gruss P: *Pax6*, a murine paired box gene, is expressed in the developing CNS. Development 1991;113:1435–1449.
41 Thomas PQ, Johnson BV, Rathjen J, Rathjen PD: Sequence, genomic organization, and expression of the novel homeobox gene *Hesx1*. J Biol Chem 1995;270:3869–3875.
42 Hermesz E, Mackem S, Mahon KA: *Rpx*: A novel anterior-restricted homeobox gene progressively activated in the prechordal plate, anterior neural plate and Rathke's pouch of the mouse embryo. Development 1996;122:41–52.

43 Zhadanov AB, Bertuzzi S, Taira M, Dawid IB, Westphal H: Expression pattern of the murine LIM class homeobox gene Lhx3 in subsets of neural and neuroendocrine tissues. Dev Dyn 1995;202:354–364.
44 Bach I, Rhodes SJ, Pearse RV, Heinzel T, Gloss B, Scully KM, Sawchenko PE, Rosenfeld MG: *P-Lim*, a LIM homeodomain factor, is expressed during pituitary organ and cell commitment and synergizes with *Pit1*. Proc Natl Acad Sci USA 1995;92:2720–2724.
45 Sheng HZ, Moriyama K, Yamashita T, Li H, Potter SS, Mahon KA, Westphal H: Multistep control of pituitary organogenesis. Science 1997;278:1809–1812.
46 Sheng HZ, Zhadanov AB, Mosinger B Jr, Fujii T, Bertuzzi S, Grinberg A, Lee EJ, Huang SP, Mahon KA, Westphal H: Specification of pituitary cell lineages by the LIM homeobox gene *Lhx3*. Science 1996;272:1004–1007.
47 Lamonerie T, Tremblay JJ, Lanctot C, Therrien Y, Gauthier Y, Drouin J: Ptx1, a *bicoid*-related homeo box transcription factor involved in transcription of the proepiomelanocortin gene. Genes Dev 1996;10:1284–1295.
48 Szeto DP, Ryan AK, O'Connell SM, Rosenfeld MG: P-OTX: A Pit1 interacting homeodomain factor expressed during anterior pituitary gland development. Proc Natl Acad Sci USA 1996; 93:7706–7710.
49 Simmons DM, Voss JW, Ingraham HA, Holloway JM, Broide RS, Rosenfeld MG, Swanson LW: Pituitary cell phenotypes involve cell-specific Pit-1 mRNA translation and synergistic interactions with other classes of transcription factors. Genes Dev 1990;4:695–711.
50 Japon MA, Rubinstein M, Low MJ: In situ hybridization analysis of anterior pituitary hormone gene expression during fetal mouse development. J Histochem Cytochem 1994;42: 1117–1125.
51 Ingraham HA, Lala DS, Ikeda Y, Luo X, Shen WH, Nachtigal MW, Abbud R, Nilson JH, Parker KL. The nuclear receptor steroidogenic factor-1 acts at multiple levels of the reproductive axis. Genes Dev 1994;8:2302–2312.
52 Li S, Crenshaw EBI, Rawson EJ, Simmons DM, Swanson LW, Rosenfeld MG: Dwarf locus mutants which lack three pituitary cell types result from mutations in the POU domain gene Pit1. Nature 1990;347:528–533.
53 Sornson MW, Wu W, Dasen JS, Flynn SE, Norman DJ, O'Connell SM, Gukovsky I, Carriere C, Ryan AK, Miller AP, Zuo L, Gleiberman AS, Andersen B, Beamer WG, Rosenfeld MG: Pituitary lineage determination by the Prophet of Pit-1 homeodomain factor defective in Ames dwarfism. Nature 1996;384:327–333.
54 Thomas P, Beddington RSP: Anterior primitive endoderm may be responsible for patterning the anterior neural plate in the mouse embryo. Curr Biol 1996;6:1487–1496.
55 Thomas P, Brickman J, Popperl H, Krumlauf R, Beddington RSP: Axis duplication and anterior identity in the mouse embryo; in: Symposia on Quantitative Biology. New York, Cold Spring Harbor Laboratory Press, 1997, vol LXII, pp 115–125.
56 Wilson DS, Guenther B, Desplan C, Kuriyan J: High resolution crystal structure of a paired (Pax) class cooperative homeodomain dimer on DNA. Cell 1995;82:709–719.
57 Kissinger CR, Liu BS, Martin-Blanco E, Kornberg TB, Pabo CO: Crystal structure of an engrailed homeodomain-DNA complex at 2.8 Å resolution: A framework for understanding homeodomain-DNA interactions. Cell 1990; 63:579–590.
58 Okamoto N, Wada Y, Ida S, Koga R, Ozono K, Chiyo H, Hayashi A, Tatsumi K: Monoallelic expression of normal mRNA in the *PIT1* mutation heterozygotes with normal phenotype and biallelic expression in the abnormal phenotype. Hum Mol Genet 1994;3:1565–1568.

Hormone Research

Horm Res 2000;53(suppl 1):34–37

How Can Molecular Biology Contribute to the Management of Congenital Adrenal Hyperplasia?

E. Martin Ritzén[a] Svetlana Lajic[a,b] Anna Wedell[b]

Departments of [a]Woman and Child Health and [b]Molecular Medicine, Karolinska Institute, Stockholm, Sweden

Key Words

Genotype · Glucocorticoids · 21-Hydroxylase deficiency · Phenotype · Prenatal treatment

Abstract

The most common form of congenital adrenal hyperplasia is due to a deficiency of 21-hydroxylase (21OHD) activity and is caused by a mutation in the CYP21 gene. By genotyping patients, new and important information can be gained, including presence or absence of 21OHD in borderline cases, determining the severity of disease and identifying heterozygote carriers. Current management of patients with 21OHD involves administering sufficient glucocorticoids to suppress excess adrenal androgen secretion, but not so much that bone growth and mineralization are impaired. New management strategies have been proposed and include administering only substitution doses of corticosteroids and counteracting side-effects by administering an anti-androgen and aromatase inhibitor. Adrenalectomy has also been proposed. Further investigation into these approaches is necessary.

Introduction

Congenital adrenal hyperplasia due to a total or partial deficiency in 21-hydroxylase (21OHD) activity is prevalent all over the world. In almost all cases, 21OHD has been shown to be due to deletion or a point mutation in the CYP21 gene (*Cy*tochrome *P*450c*21*), causing a complete (null) or partial lack of enzyme activity. The traditional categorization of 21OHD patients into 'salt losing' (those that show neonatal hyponatraemia and hyperkalaemia), 'simple virilizing' (where affected girls are virilized and boys develop precocious puberty, but maintain normal electrolyte balance under baseline conditions) and 'non-classical' forms (with late onset) seems to be too simplified in the light of new understanding of the variable function of the 21OH – there are no absolute borders between the different degrees of severity. A 'simple virilizer' may develop hypernatraemia when under stress and thus become a 'salt loser'. When it is possible to determine genotypes objectively, this may be a better base for categorization, provided that a relation between genotype and phenotype can be established.

KARGER

Fax +41 61 306 12 34
E-Mail karger@karger.ch
www.karger.com

0301–0163/00/0537–0034$17.50/0

Accessible online at:
www.karger.com/journals/hrc

Prof. E.M. Ritzén
Astrid Lindgren Children's Hospital, Q2:08
Karolinska Hospital
S–171 76 Stockholm (Sweden)
Tel. +46 8 51772465, Fax +46 8 51775128, E-Mail martin.ritzen@kbh.ki.se

Genotype/Phenotype Relations in 21OHD

By using PCR and, if necessary, sequencing methods, it has been possible to define the genotype of a large number of patients and compare this with the clinical presentation. It has been noticed that the most severe (null) mutations always cause a severe form of 21OHD, with early presentation of virilization, dehydration, salt loss and vomiting, while other mutations are associated with a milder form of the disease, presenting after the neonatal period [1–3]. However, there are a few genotypes that may show a somewhat variable presentation. A splice mutation in intron 2 (I2splice), which occurs in 25–30% of all mutations, will severely impair enzyme function. In most patients this causes a severe, salt-losing form, while a few may manage normal electrolyte homeostasis under baseline conditions [1]. Likewise, most patients with the point mutation I172N have virilization but normal salt balance, while a few may show salt loss. Also, occasional families have been described where there is a marked deviation between phenotypes of siblings with apparently the same genotype [2].

When comparing the phenotype resulting from the mutations found in a large number of patients, the general rule was that the best functioning allele determined the severity of the disease [1]. However, there is also a contribution from the other allele, such that a compound heterozygote with one mildly and one severely affected gene will generally have a more severe phenotype than one who carries two 'mild' alleles.

Practical Use of Mutation Analysis

For most patients in whom 21OHD is suspected, the diagnosis can be confirmed or rejected by a simple determination of 17-hydroxyprogesterone levels in blood, with or without prior stimulation with adrenocorticotrophic hormone. Thus, mutation analysis may seem to be expensive and unnecessary. However, in several instances new and clinically important information can be gained by genotyping the patient:

In *prenatal diagnosis*, where the presence or absence of 21OHD can be readily ascertained by mutation analysis of DNA from a chorionic villous biopsy at week 10 of gestation. This is important both for genetic counselling and for monitoring of prenatal treatment of 21OHD (see below).

In *neonatal screening programmes* a single determination of blood 17-hydroxyprogesterone may reveal the diagnosis, but in boys it is not possible to determine the severity of the disease in this way. Mutation analysis in the newborn infant, however, has proven to be strongly related to the clinical form [4]. This may have therapeutic implications, as it seems that only low doses of glucocorticoids are needed in moderately severe forms of 21OHD [5].

Identification of heterozygote carriers: Provided that the index case and the parents have been genotyped, the carrier status of siblings can be accurately determined by mutation analysis, but not with certainty by hormone assays. However, proving absence of a mutation by PCR techniques may be misleading, if parents have not been studied. PCR alone will not identify a carrier with one normal allele and one allele with a total deletion of CYP21.

In *borderline cases,* when hormone assays in blood or urine have failed to establish whether 21OHD is present or not, mutation analysis can provide useful information.

When *glucocorticoid treatment* has been started before adequate blood sampling has been done, mutation analysis will confirm or disprove the diagnosis without interfering with the medical treatment.

In *rejecting prenatal treatment,* when the index case is a boy with undefined severity of 21OHD. Only fetuses with severely impaired function of 21OH should be considered for treatment, and mutation analysis may be the only way to distinguish severe from moderate forms in the older brother.

Prenatal Treatment of 21OHD

Girls with severe forms of 21OHD are born virilized, often to the extent that the sex of the newborn infant is uncertain. This is a psychological burden for the parents which might influence their attitude to the child for years to come. Therefore, it was highly acclaimed by both parents and doctors when it was shown that most, if not all, of the virilization could be prevented if the mother was treated with dexamethasone from week 7 to 8 of gestation [6]. Later, caution was advised because exposure of the fetus to a potent glucocorticoid like dexamethasone might be hazardous to the child and cause long-term side-effects [7–9]. The situation is particularly worrying as all babies at risk must be treated from week 7 of gestation until the female sex and the diagnosis can be confirmed (in 1 child out of 8) or disproved (in 7 children out of 8). The latter are treated unnecessarily during weeks 7–11.

In order to establish whether unwanted effects of the dexamethasone treatment could be seen on mother or child, a controlled study of all 44 children born after such treatment in Sweden and Norway was carried out [10]. For each child treated, 2 healthy children with the same sex and birth date and from the same maternity unit were recruited as controls.

Compared with the controls, the dexamethasone-treated mothers complained more about feeling 'swollen' and irritated during the pregnancy, and they gained weight significantly more than controls during the 1st trimester, but not during later pregnancy. Only one parameter was better in the dexamethasone-treated group: there was no case of pain from 'symphysiolysis' in the dexamethasone group, while 16% of the control mothers had this problem.

Only 5 girls were diagnosed as having 21OHD at the chorionic villous biopsy and were treated until term. All had minor virilization as compared with their elder sisters (1 needed surgery for labial fusion). Three of the 5 had some adverse events. One was the product of a 37-week gestation that ended in caesarean section due to pre-eclampsia. She was initially –4 standard deviation scores in length, but later recovered in all respects. Two affected sisters failed to thrive, 1 of whom had a concurrent mitochondrial disease. In the short-term-treated group, one unaffected boy was born with hydrocephalus and agenesis of the corpus callosum. Another whose parents were first cousins had unexplained ataxia and mental retardation at age 3 years. Three unaffected girls also had some adverse medical events: 1 had initial failure to thrive, but fully recovered later (adenovirus infection was suspected), 1 had unexplained poor growth, and 1 girl had bouts of mood fluctuations and aggressiveness. In the control group, there was 1 case of Down's syndrome, but otherwise no record of adverse events.

In this study, only growth parameters could be followed quantitatively during childhood. There were no significant differences in length/height, weight or head circumference between the healthy children treated during weeks 7–11 of gestation and the matched controls at 3 months, 1 year and 4 years of age. Thus, prenatal treatment during the 1st trimester does not seem to influence the pre- or postnatal growth of the children. The 5 affected girls treated throughout gestation are too few to allow statistical evaluation.

The conclusion of the above study is that prenatal dexamethasone treatment should still be considered 'a promising experimental procedure of unproven safety' [8]. Such treatment should, therefore, be organized as planned clinical studies, so that solid data on its safety can be accumulated as soon as possible.

Results of Present Treatment Regimens in 21OHD, as Reflected in Adulthood

Follow-up of body composition of patients with 21OHD into adulthood reflects both treatment during childhood and adolescence (as shown in final height) and in adulthood (weight for height, body fat content and bone mineralization). Furthermore, studies of quality of life as measured by partnership, parenthood, educational and professional status, employment situation and others may reveal important aspects on both former treatment in childhood and later as adults. Such studies are scarce. Common to all reports, however, is short stature and often some degree of obesity [3, 11–13]. This could be interpreted as sequels of less than optimal management during childhood and adolescence, though in the Finnish study, those diagnosed at an early age achieved greater adult height than those with late diagnosis [14].

Possible New Strategies for Treatment of Patients with 21OHD

The medical management of patients with 21OHD has been largely unchanged over the last decades: administer sufficient glucocorticoids to suppress excess adrenal androgen secretion, but not so much that obesity develops, bone growth or mineralization is impaired. This balance is difficult to achieve – many patients end up shorter and fatter than their genetic potential. Some authors have questioned whether it is at all possible to suppress adrenal androgens completely without giving excess glucocorticoids with their accompanying side-effects [15].

Two opposite new strategies have been proposed to overcome the shortcomings of present management. One is to give not more than normal substitution doses of glucocorticoids, thus accepting that excess adrenal androgen is produced, but counteracting the unwanted effects of these androgens by simultaneous administration of an anti-androgen and an aromatase inhibitor (the latter to suppress androgen metabolism to oestrogens which are supposedly the cause of premature bone maturation) [15]. Short-term results of this approach seem promising [16]. Another approach for patients who have only non-functioning CYP21 genes is to remove the adrenal glands and treat the patients with physiological (i.e., lower than presently used) doses of glucocorticoids, as in Addison's disease [15]. The reasoning behind this seemingly drastic measure is that in the absence of functioning CYP21 genes, no gluco- or mineralocorticoids can be produced. Rather, the excess precursors (mainly 17-hydroxyprogesterone) in stress situations will inhibit the effect of administered mineralocorticoids. The latter phenomenon has been experimentally shown in patients with congenital adrenal hyperplasia [17] and was recently confirmed in 2 patients with null alleles for CYP21 prior to adrenalectomy [18; Ritzén, unpubl. data]. The question remains whether ablation of the adrenal medulla will have any adverse effects. The total concentration of catecholamines in blood normalizes soon after adrenalectomy [19], probably due to increased activity in paraganglia and sympathetic nerve endings. However, there may be an increased risk that in the absence of adrenaline 'stress symptoms' (tachycardia, paleness, sweating, hypoglycaemia) might pass unnoticed for a longer time, if glucocorticoid doses are not adequately increased during stress. Also, the absence of the adrenaline counteraction of hypoglycaemia might not be fully compensated by noradrenaline from other sources. Two of the 3 patients with 21OHD who have been adrenalectomized in an ongoing prospective clinical study have had sudden hypoglycaemic episodes with convulsions during gastro-enteritis. These few observations suggest that parents and patients should be advised to be extra careful and be sure to increase hydrocortisone doses during stress, if the child has been adrenalectomized.

References

1 Wedell A, Thilén A, Ritzén EM, Stengler B, Luthman H: Mutational spectrum of the steroid 21-hydroxylase gene in Sweden: Implications for genetic diagnosis and association with disease manifestation. J Clin Endocrinol Metab 1994;78:1145–1152.

2 Wilson RC, Mercado AB, Cheng KC, New MI: Steroid 21-hydroxylase deficiency: Genotype may not predict phenotype. J Clin Endocrinol Metab 1995;80:2322–2329.

3 Jääskeläinen J, Levo A, Voutilainen R, Partanen J: Population-wide evaluation of disease manifestation in relation to molecular genotype in steroid 21-hydroxylase (CYP21) deficiency: Good correlation in a well defined population. J Clin Endocrinol Metab 1997;82: 3293–3297.

4 Nordenström A, Thilén A, Hagenfeldt L, Larsson A, Wedell A: Genotyping is a valuable diagnostic complement to neonatal screening for congenital adrenal hyperplasia due to steroid 21-hydroxylase deficiency. J Clin Endocrinol Metab 1999;84:1505–1509.

5 Thilén A, Woods KA, Perry LA, Savage MO, Wedell A, Ritzén M: Early growth is not increased in untreated moderately severe 21-hydroxylase deficiency. Acta Paediatr 1995;84: 894–898.

6 David M, Forest MG: Prenatal treatment of congenital adrenal hyperplasia resulting from 21-hydroxylase deficiency. J Pediatr 1984;105: 799–803.

7 Forest MG, Morel Y, David M: Prenatal treatment of congenital adrenal hyperplasia. Trends Endocrinol Metab 1998;9:284–289.

8 Miller WL: Prenatal treatment of congenital adrenal hyperplasia: A promising experimental procedure of unproven safety. Trends Endocrinol Metab 1998;9:290–293.

9 Ritzén EM: Prenatal treatment of congenital adrenal hyperplasia: A commentary. Trends Endocrinol Metab 1998;9:293–295.

10 Lajic S, Wedell A, Bui T-H, Ritzén EM, Holst M: Long-term somatic follow-up of prenatally treated children with congenital adrenal hyperplasia. J Clin Endocrinol Metab 1998;83:3872–3880.

11 Urban M, Peter L, Migeon C: Adult height and fertility in men with congenital virilising adrenal hyperplasia. N Engl J Med 1978;299:1392–1396.

12 Helleday J, Siwers B, Ritzén EM, Carlström K: Subnormal androgen and elevated progesterone levels in women treated for congenital virilising 21-hydroxylase deficiency. J Clin Endocrinol Metab 1993;76:933–936.

13 Girgis R, Winter JS: The effects of glucocorticoid replacement therapy on growth, bone mineral density, and bone turnover markers in children with congenital adrenal hyperplasia. J Clin Endocrinol Metab 1997;82:3926–3929.

14 Jääskeläinen J, Voutilainen R: Growth of patients with 21-hydroxylase deficiency: An analysis of the factors influencing adult height. Pediatr Res 1997;40:30–33.

15 Van Wyk JJ, Gunther DF, Ritzén EM, Wedell A, Cutler GB Jr, Migeon CJ, New MI: The use of adrenalectomy as a treatment for congenital adrenal hyperplasia. J Clin Endocrinol Metab 1996;81:3180–3190.

16 Laue L, Merke DP, Jones JV, Barnes KM, Hill S, Cutler GB Jr: A preliminary study of flutamide, testolactone, and reduced hydrocortisone dose in the treatment of congenital adrenal hyperplasia. J Clin Endocrinol Metab 1996;81: 3535–3539.

17 Janoski AH: Naturally occurring adrenal steroids with salt-losing properties: Relationship to congenital adrenal hyperplasia; in Lee PA, Plotnick LP, Kowarski AA, Migeon CJ (eds): Congenital Adrenal Hyperplasia. Baltimore, University Park Press, 1977, pp 99–112.

18 Gunther DF, Bukowski TP, Ritzén EM, Wedell A, Van Wyk JJ: Prophylactic adrenalectomy of a three-year old girl with congenital adrenal hyperplasia: Pre- and postoperative studies. J Clin Endocrinol Metab 1997;82:3324–3327.

19 Hamberger B, Telenius-Berg M, Cedermark B, Gröndal S, Hansson BG, Werner S: Subtotal adrenalectomy in multiple endocrine neoplasia type 2. Henry Ford Hosp Med J 1987;35:127–128.

Hormone Research

Horm Res 2000;53(suppl 1):38–41

Neonatal Endocrinology of Abnormal Male Sexual Differentiation: Molecular Aspects

Olaf Hiort

Department of Paediatrics, Medical University of Lübeck, Germany

Key Words

Sexual differentiation · Gonadal development · Androgen formation · Genetics

Abstract

Normal male sexual differentiation is a complex mechanism, depending on genetic and hormonal control. The bipotent gonad arises at the genital ridge under the control of autosomal genes which are also involved in the formation of other organs. Progression towards testicular differentiation is mediated through both autosomal and gonosomal genes, leading to alignment of Sertoli cells and Leydig cells. Within the latter, androgen formation is induced by human chorionic gonadotrophin and luteinizing hormone. The influence of testosterone and dihydrotestosterone leads to differentiation of epidydimis, vasa deferentia, prostate and external genitalia. Androgens mediate their action through the androgen receptor, a nuclear transcription factor controlling the regulation of so far unknown target genes. Genetic variations within this pathway interrupt normal male development and will lead to malformation of external and/or internal genitalia.

Genetic Control of Gonadal Formation

Genetic control of male sexual differentiation begins at conception with chromosomal organization. However, during early embryogenesis, the gonad is still indifferent and can potentially differentiate into either ovary or testis. The development of the indifferent gonad is induced in the genital ridge under the control of autosomal genes acting as transcription factors (table 1). These factors are generally also involved in the development of other organs. These genes include *WT-1* and *LIM-1*. Both genes are also important for renal besides gonadal development; however, naturally occurring mutations in humans have only been described in *WT-1*. Mutations in *WT-1* are associated with syndromes involving gonadal dysgenesis, development of gonadoblastoma (Frasier syndrome), nephropathy and Wilms' tumours (Denys-Drash syndrome), anomalies of the eye and mental retardation (WAGR syndrome) [1]. Of significance for both gonadal and adrenal development is the steroidogenic factor 1, encoded by the *FTZF1* gene. This factor is an orphan nuclear receptor harbouring a two zinc finger DNA-binding domain. It is regulating several genes important both to gonadal and adrenal development as well as steroidogenesis [2].

KARGER

Fax +41 61 306 12 34
E-Mail karger@karger.ch
www.karger.com

0301–0163/00/0537–0038$17.50/0

Accessible online at:
www.karger.com/journals/hre

Dr. O. Hiort
Department of Paediatrics, Medical University of Lübeck
Ratzeburger Allee 160
D–23538 Lübeck (Germany)
Tel. +49 451 500 2548, Fax +49 451 500 2184, E-Mail hiort@paedia.mu-luebeck.de

Table 1. Chromosomal localization of genes involved in gonadal determination and associated malformations in the event of genetic variation apart from gonadal dysgenesis

Name	Chromosome location	Human disorder	Associated malformation
WT-1	11p13	yes	Wilms' tumours, nephropathy, gonadoblastoma
LIM-1	11p12–13	no	–
SF-1 *(FTZF1)*	9q33	yes	Adrenal insufficiency
SRY	Yp11.3	yes	none
DAX-1	Xp21	yes	(duplication – sex reversal) (mutation – adrenal insufficiency)
SOX-9	17q24.3–25.1	yes	campomelic dysplasia
DMRT1/DMRT2	9p24.3	yes	renal malformation, mental retardation

Disruption of the genetic cascade with inhibition of the formation of the indifferent gonad can potentially occur in either sex; however, only in males will sex reversal take place due to subsequent failure of androgen formation. In contrast, several genes have been characterized that are involved specifically in the progression of the indifferent gonad toward testicular tissue. At about 6–7 weeks of gestation, the alignment of Sertoli cells occurs within the seminiferous cords. The Sertoli cells in turn give rise to the differentiation of the Leydig cells at about week 8 of gestation. The first gene to be known in this differentiation process was *SRY*, the sex-determining region of the Y chromosome [3]. *SRY* is localized on the short arm of the Y chromosome and encompasses a single exon gene with a DNA-binding domain. Its counterpart is the *DAX-1* gene on the X chromosome. *DAX-1* itself is not crucial for testicular development, as disruptive mutations of *DAX-1* do not inhibit testicular formation but rather lead to adrenal hypoplasia and hypogonadotrophic hypogonadism [4]. However, duplication of the chromosomal region on Xp21 harbouring *DAX-1* leads to a dosage-sensitive sex reversal in males, probably due to repression of *SRY*. This suggests a role of *DAX-1* as a repressor of testicular formation. *DAX-1* may also act as a repressor on *SOX-9*, an autosomal gene involved in both gonadal and cartilage formation. Mutations in *SOX-9* lead to campomelic dysplasia and sex reversal in affected karyotypic males [5].

In addition to these known genes of gonadal development, a number of genes are suspected to play an important role as sexual regulators of testicular differentiation. Deletions of chromosomes 9p and 10q have been associated with sex reversal in 46,XY individuals. Most of these patients also exhibit additional phenotypic features, such as mental retardation and craniofacial abnormalities [6]. The degree of genital ambiguity may be variable. Of special interest is a small region on chromosome 9p23–24. This region harbours two genes of interest for testicular development which are both deleted in a hemizygous concomitant fashion in 46,XY sex reversal. These genes, named *DMRT1* and *DMRT2*, are related to proven sex-determining genes in other organisms and in humans are solely expressed in the testes [7].

Genetic Control of Androgen Biosynthesis

The differentiation of steroid-forming Leydig cells is under control of gonadotrophins (fig. 1). During late gestation, placental human chorionic gonadotrophin is replaced as the primary controlling hormone by the luteinizing hormone of the fetal pituitary gland. Testosterone is synthesized in the Leydig cells from cholesterol via five enzymatic steps from about week 8 of gestation. A pivotal role of steroidogenesis is the influx of cholesterol to the inner mitochondrial membrane to be processed by the P450 enzyme complex. This rate-limiting step of steroid synthesis is mediated in both gonads and adrenals by the steroidogenic acute regulatory (StAR) protein. Mutations of StAR protein and disruption of the active mitochondrial transport lead to congenital adrenal lipoid hyperplasia and to gonadal failure in karyotypic males with sex reversal [8]. In contrast, genetic variations in the P450 side-chain cleavage enzyme catalyzing the formation of pregnenolone from cholesterol have not been described in humans. These fetuses are probably not viable due to the importance of this enzyme in placental steroid formation which is independent of StAR. In all other enzymes involved in androgen biosynthesis, genetic defects have been characterized in humans. If these mutations occur in enzymes involved in both adrenal and gonadal steroid formation, accompanying congenital adrenal hyperplasia will occur. In males, defective virilization may be due to

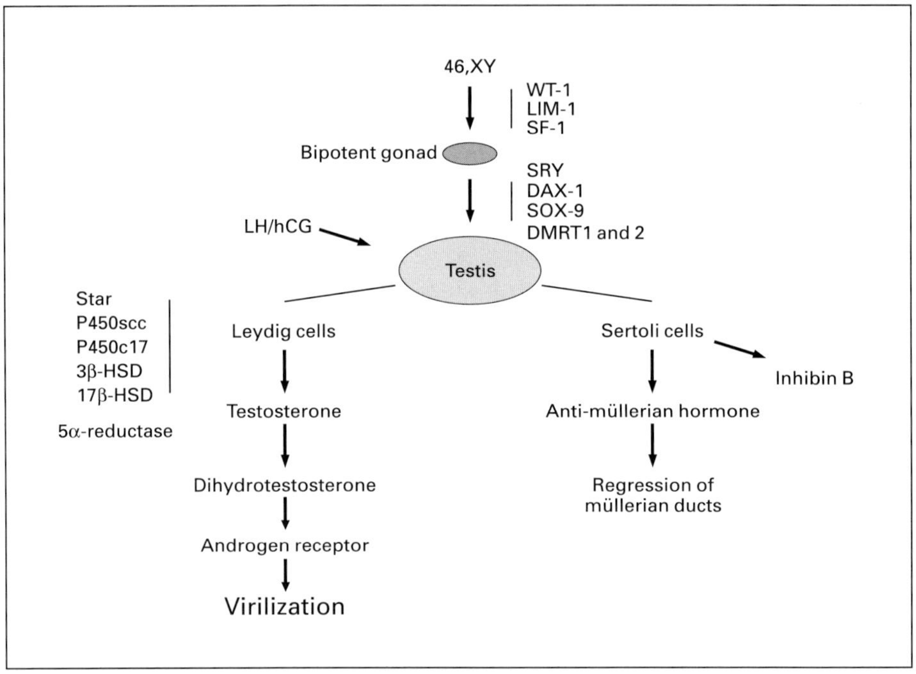

Fig. 1. Pathways of normal male sexual differentiation. Genetic control of development of the bipotent gonad and testicular differentiation lead to hormonal control of the development of the internal and external genitalia.

3β-hydroxysteroid dehydrogenase deficiency caused by mutations in the gene encoding the type II enzyme. Also mutations in the P450c17 enzyme complex may be the basis for combined virilization deficit and congenital adrenal hyperplasia in males [9].

Isolated diminished androgen synthesis has rarely been reported for mutations in the P450c17 enzyme complex affecting only 17/20-lyase activity [10]. More important in defective virilization are genetic variations in the 17β-hydroxysteroid dehydrogenase (17β HSD) [11]. Of the five currently known isoenzymes, the type 3 enzyme converts androstenedione to testosterone within the Leydig cells. To date, 17 different mutations have been described within the 17β-HSD 3 gene [12]. These patients are usually born with a completely or almost completely female phenotype, no müllerian structures and normal infantile testes frequently located in the inguinal canal. Although severe testosterone deficiency has occurred, astonishingly, epididymal structures are normally developed. At the time of expected puberty, rather striking virilization is noticed, most probably due to testosterone synthesis via other isoenzymes of 17β-HSD. A similar phenotype is observed in patients carrying mutations in the 5α-reductase type 2 which converts testosterone to dihydrotestosterone within the peripheral androgen target tissues. In addition, these affected 46,XY children are often born with a female phenotype, but demonstrate virilization at the expected time of puberty. This may be due to rising testosterone levels or increasing activity of the isoenzyme 5α-reductase type 1 [13].

Molecular Events of Androgen Action

Androgens exert their specific action via the androgen receptor (AR) which is also a nuclear transcription factor. The AR is expressed from week 9 of gestation in embryos of both sexes, hence excessive amounts of androgens in 46,XX embryos also lead to unwanted virilization via the AR. Interestingly, the AR gene is localized on the X-chromosome, thus, 46,XY individuals carry only one copy of this gene. A large number of mutations have been described within the AR gene leading to a highly variable phenotype of virilization deficit ranging from an unequivocal female phenotype over ambiguous genitalia to males with only slightly defective virilization associated with abnormal sperm maturation within the testis [14]. Knowledge of predictors of the phenotype due to a specific genetic abnormality within the AR gene is limited. Phenotypic variability may be due to alternating androgen levels during early fetal life, changing the susceptibility of the mutated AR to induce target gene activity [15]. In a significant proportion of patients with a mutated AR gene, the genetic aberration has occurred after conception as an early somatic mutation [16, 17]. This genetic pattern, also observed in other X-chromosomal disorders, leads to a mosaicism of mutated and wild-type AR proteins and, hence, to increased androgen sensitivity. Whether alternative pathways of androgen action are significant in the determination of the male phenotype remains questionable at this time [18].

Acknowledgements

This work was supported by grants from the Deutsche Forschungsgemeinschaft (Hi 497/3-3 and 4-2) and the Forschungsförderungsprogramm des Universitätsklinikums Lübeck.

References

1 Hughes IA, Coleman N, Faisal Ahmed S, Ng KL, Cheng A, Lim HN, Hawkins JR: Sexual dimorphism in the neonatal gonad. Acta Paediatr 1999;88(suppl 428):23–30.

2 Achermann JC, Ito M, Ito M, Hindmarsh PC, Jameson JL: A mutation in the gene encoding steroidogenic factor-1 causes XY sex reversal and adrenal failure in humans. Nat Genet 1999;22:125–126.

3 Capel B: Sex in the 90s: SRY and the switch to male pathway. Annu Rev Physiol 1998;60: 497–523.

4 Yu RN, Ito M, Saunders TL, Camper SA, Jameson JL: Role of Ahch in gonadal development and gametogenesis. Nat Genet 1998;20: 353–357.

5 Lim HN, Hawkins JR: Genetic control of gonadal differentiation. Baillières Clin Endocrinol Metab 1998;12:1–16.

6 Pfeiffer RA, Rauch A, Trautmann U, Dörr HG, Hiort O, Scherer G, Rösch G, Papadopoulos T, von der Hardt K, Lachmann E: Defective sexual development in an infant with 46,XY, der(9)t(8;9)(q23.1;p23)mat. Eur J Pediatr 1999;158:213–216.

7 Raymond CS, Parker ED, Kettlewell JR, Brown LG, Page DC, Kusz K, Jaruzelska J, Reinberg Y, Flejter WL, Bardwell VJ, Hirsch B, Zarkower D: A region of human chromosome 9p required for testis development contains two genes related to known sexual regulators. Hum Mol Genet 1999;8:989–996.

8 Stocco DM: An update on the mechanism of action of the steroidogenic acute regulatory (StAR) protein. Exp Clin Endocrinol Diabetes 1999;107:229–235.

9 Miller WL: Early steps in androgen biosynthesis: From cholesterol to DHEA. Baillières Clin Endocrinol Metab 1998;12:67–81.

10 Geller DH, Auchus RJ, Mendonca BB, Miller WL: The genetic and functional basis of isolated 17,20-lyase deficiency. Nat Genet 1997; 17:201–205.

11 Andersson S, Geissler WM, Wu L, Davis DL, Grumbach MM, New MI, Schwarz HP, Blethen SL, Mendonce BB, Bloise W, Witchel SF, Cutler GB, Griffin JE, Wilson JD, Russell DW: Molecular genetics and pathophysiology of 17β-hydroxysteroid dehydrogenase 3 deficiency. J Clin Endocrinol Metab 1996;81:130–136.

12 Twesten W, Holterhus PM, Sippell WG, von Mühlendahl KE, Schumacher H, Morlot M, Schenk B, Hiort O. Klinische, endokrinologische und molekulargenetische Befunde bei 17β-Hydroxysteroid-Dehydrogenase (17β-HSD)-Mangel. Monatsschr Kinderheilkd 1999;147(suppl 2):85.

13 Hiort O, Willenbring H, Albers N, Hecker W, Engert J, Dibbelt L, Sinnecker GHG: Molecular genetic analysis and hCG stimulation tests in the diagnosis of prepubertal patients with partial 5α-reductase deficiency. Eur J Pediatr 1996;155:445–451.

14 Hiort O, Sinnecker GHG, Holterhus PM, Nitsche EM, Kruse K: The clinical and molecular spectrum of androgen insensitivity. Am J Med Genet 1996;63:218–222.

15 Holterhus PM, Homburg N, Struve D, Hiort O: Partial activity of mutant Leu712Phe androgen receptor – a likely basis for marked phenotypic variability in a PAIS family. Horm Res 1999;51(suppl 2):75.

16 Hiort O, Sinnecker GHG, Holterhus PM, Nitsche EM, Kruse K: Inherited and de novo androgen receptor gene mutations: Investigation of single-case families. J Pediatr 1998;132: 939–943.

17 Holterhus PM, Wiebel J, Sinnecker GHG, Brüggenwirth HAT, Sippell WG, Brinkmann AO, Kruse K, Hiort O: Clinical and molecular spectrum of somatic mosaicism in androgen insensitivity syndrome. Pediatr Res 1999;46: 684–690.

18 Brinkmann AO, Blok LJ, de Ruiter PE, Doesburg P, Steketee K, Berrevoets CA, Trapman J: Mechanisms of androgen receptor activation and function. J Steroid Biochem Mol Biol 1999;69:307–313.

Hormone Research

Horm Res 2000;53(suppl 1):42–49

Neonatal and Post-Natal Growth

Alan T. Gibson[a] Sally Carney[b] Elena Cavazzoni[a] Jeremy K.H. Wales[b]

[a]Jessop Hospital for Women, Sheffield, and [b]Division of Paediatrics, University of Sheffield, UK

Key Words
Growth, measurement · Intra-uterine growth retardation · Small for gestational age · Pre-term infants · Outcome

Abstract
Growth velocity is higher in late intra-uterine and early post-natal life than at any time thereafter, and accurate measurements are essential for appropriate monitoring. The accuracy with which such measurements are made and recorded is frequently questionable, however, and short- and medium-term changes in growth may be difficult to interpret in the light of normal variations in the pattern of growth. Infants who are small at birth must be accurately classified because intra-uterine growth retardation and small for gestational age have different implications for both causation and outcome. Prediction of expected growth on the basis of mid-parental height is essential but frequently omitted. Post-natal growth impairment is common in pre-term infants and is often rapid in onset. Poor growth may continue for many months, and catch-up may be incomplete. Early growth failure may have a significant influence on subsequent morbidity and mortality.

Introduction

Since the detailed measurements of Count de Montbeillard in the 18th century many studies have confirmed that growth velocity is higher at birth than at any point thereafter. Although early post-natal growth velocity is high, it is far exceeded by that of the developing fetus. In the early stage of intra-uterine life the growth rate is relatively slow as embryonic differentiation occurs, but rapidly increases from the 10th post-menstrual week to reach a peak approximately half way through gestation. At this point it is more than three times as rapid as it will be at term [1].

Measuring and Plotting

Accurate measurements of growth in the post-natal period are essential for accurate monitoring, but although such measurements are commonly made, accuracy is often questionable. This is compounded in premature infants where small size and restrictions imposed by intensive care often result in infrequent measurements of weight alone. We have assessed the accuracy of measurements made at birth by re-measuring 250 infants who had been admitted to the post-natal wards or the neonatal unit

KARGER

Fax + 41 61 306 12 34
E-Mail karger@karger.ch
www.karger.com

0301–0163/00/0537–0042$17.50/0

Accessible online at:
www.karger.com/journals/hre

Dr. A.T. Gibson
Neonatal Intensive Care Unit
Jessop Hospital for Women, Leavygreave Road
Sheffield S3 7RE (UK)
Tel. +44 114 2268299, Fax +44 114 2268449, E-Mail A.T.Gibson@Sheffield.ac.uk

in our hospital. All repeat measurements were made by a single experienced study nurse. In 98 infants no length measurements were made. In the remaining 152 infants birth length measurements ranged between +13.8 and −9.0% of the measurement made by the study nurse. Head circumference was not measured in 94 infants and varied between +13.4 and −19.1% of the accurate measurement in the remainder. To determine the source of this inaccuracy 50 staff members who routinely admitted babies were asked to measure a plastic doll with the typical attachments of a baby receiving intensive care. Measurements of weight varied between +19 and −6% of the known weight (with one exception where the weight recorded was more than four times that of the actual doll). Measurements of length were within 25% of the accurate length (a similar error to that observed in our cohort of live infants), while measurements of head circumference fell within +1 and −0.7% of an accurate measurement. Accurate measurement is essential if growth of the newborn infant is to be monitored, yet this is frequently not achieved. Scepticism should be expressed when interpreting data where accuracy has not been validated.

Once measurements are made, they will be plotted on one of the many available growth charts, thereby allowing growth trajectory to be monitored. When the accuracy of such plots has been assessed, nearly 30% of the points were found to have been plotted incorrectly. Errors occurred along the horizontal axis in nearly 95% of the cases – frequently due to failure to correct for the degree of prematurity –, and 5% occurred along the vertical axis [2]. Many different growth charts are available, and those who use them should be aware that the same data plotted on different charts may suggest widely varying patterns of growth. If values for length gain along the 50th centile between 33 and 39 weeks derived from the Child Growth Foundation 1996/1 charts [3] are re-plotted on Tanner and Whitehouse charts [4], the line now starts below the 50th centile, drops toward the 25th centile and finishes above the 50th centile. On Gairdner and Pearson charts [5], the line starts above the 90th centile and drops to the 75th centile at 39 weeks, while on Keen and Pearse charts [6], the line starts on the 75th centile and finishes on a point midway between the 50th and 25th centile. All four charts are in use in the UK. It is important that the growth charts selected should be appropriate to the population being measured and that data are not transferred between charts unless this is unavoidable.

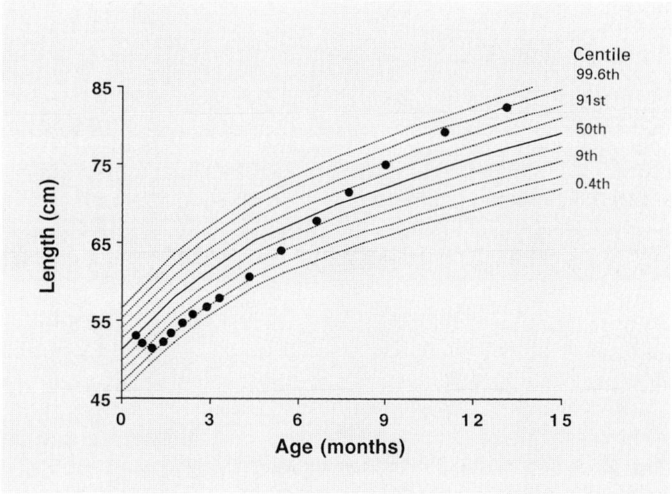

Fig. 1. Shifting centiles in a child initially referred because of concerns over poor growth in the early post-natal period. Maternal height was 155 cm, paternal height 187.5 cm.

Patterns of Growth

It is normally assumed that post-natal growth will progress smoothly along a centile line. This is not necessarily the case, as the growth trajectory may move between centile lines in the first months following birth (fig. 1). Such fluctuations may give rise to concern (as with the illustrated case), as the infant is thought to be either growing too rapidly or exhibiting growth failure. Although such changes may be attributable to significant malnutrition or clinical illness, they are often a normal occurrence, as the growth rate slowly settles to the predetermined pattern that will be followed thereafter. Some of the anxiety generated may be alleviated by the use of parental heights to establish the anticipated range for growth, yet such measurements are frequently omitted and parental stature ignored. In a study of healthy middle class American infants, shifting of growth rate was noted in two thirds of infants. In those infants where there was an upward shift in the centile charts, growth accelerated soon after birth, and a new centile was attained by a mean age of 11.5 months. Downward shift occurred later, with deceleration beginning after 3–6 months and achieving a new channel at a mean age of 13 months. Stature at birth was most closely correlated with maternal stature, while the centile to which they shifted correlated most closely with mean parental size [7].

Post-natal growth is often referred to as 'linear', but detailed measurements of short-term changes in weight

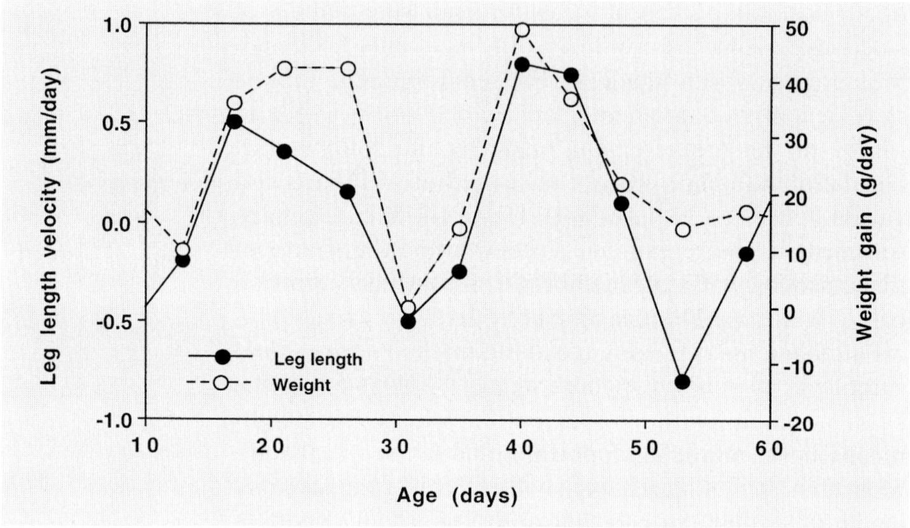

Fig. 2. Synchronized fluctuation in weight gain and skeletal growth using measurements averaged over 4- to 5-day intervals.

and leg length (using knemometry) have shown that this is not the case [8–10]. If frequent measurements are made, short-term fluctuations in both weight gain and leg growth are always seen and are often synchronous (fig. 2). These fluctuations are independent of any extraneous influences and have a periodicity that varies between infants; the interval between peak and trough varying between 10 and 20 days. It is important that observers are aware that these fluctuations occur, so that short-term increases or decreases in growth rate do not generate undue concerns.

Small for Gestational Age and Intra-Uterine Growth Retardation

Infants who are smaller than would be expected at birth present particular problems, both in classification and management. The terms 'small for gestational age' (SGA) and 'intra-uterine growth retardation' are often used indiscriminately and incorrectly and as if the definitions were interchangeable [11]. A diagnosis of intra-uterine growth retardation means that there has been evidence of growth failure during fetal development, and although measurements will have crossed centile lines, size may well remain within the normal range. A pathological process that has interfered with normal growth is implied, and infants with early sustained impairment may have a worse long-term outcome than those in whom impairment occurred later [12]. An infant is described as being SGA if birth-weight lies below a chosen centile (most commonly the 10th or 5th). This is often regarded as evidence of growth retardation, but this conclusion cannot be based upon a single measurement. If the 10th centile is used as a cut-off point, 10% of normal infants will, by definition, lie below this line. It is also important to note that fetal growth charts may well have been derived from a population with which comparison would be inappropriate.

It has been suggested that the use of an individualized birthweight ratio, where a number of confounding variables, such as gestation, maternal height, weight, parity and ethnic origin and infant's sex, are used to predict expected birthweight, may help to improve classification [13, 14]. Using this method in 616 newborn infants, 25% of those below the 10th centile for birthweight were reclassified as normally grown, while a slightly larger number of infants whose birthweight was above the 10th centile were re-classified as being inappropriately small. Reclassified infants had a similar incidence of complications during labour and delivery as those in the group to which they were re-assigned. Appropriate classification of infants with apparent fetal growth retardation is not merely a matter of semantics. Disproportionately small infants have higher mortality and morbidity in the post-natal period and are more likely to exhibit neurodevelopmental deficits in childhood [15–18]. Correct classification is, therefore, essential. Assessment of SGA infants must include measurements of both length and weight, as subsequent growth patterns and developmental outcome differ in those infants in whom both parameters are symmetrically reduced as opposed to those in whom weight is reduced but length remains within the normal range [19,

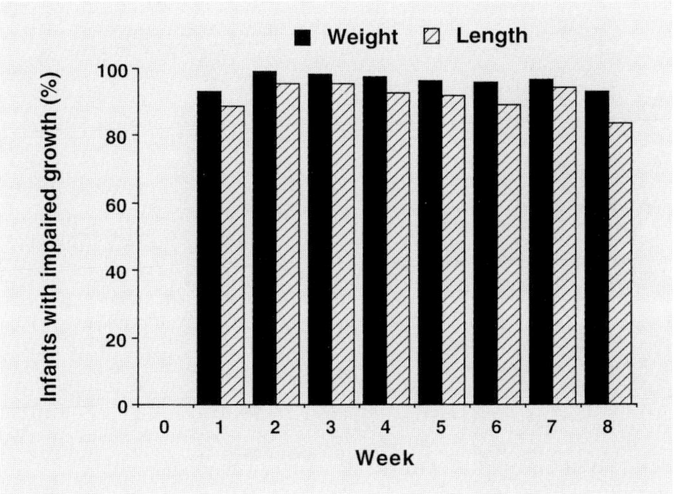

Fig. 3. Proportion of a cohort of 122 pre-term infants showing evidence of impaired weight and length gain in the first 8 post-natal weeks.

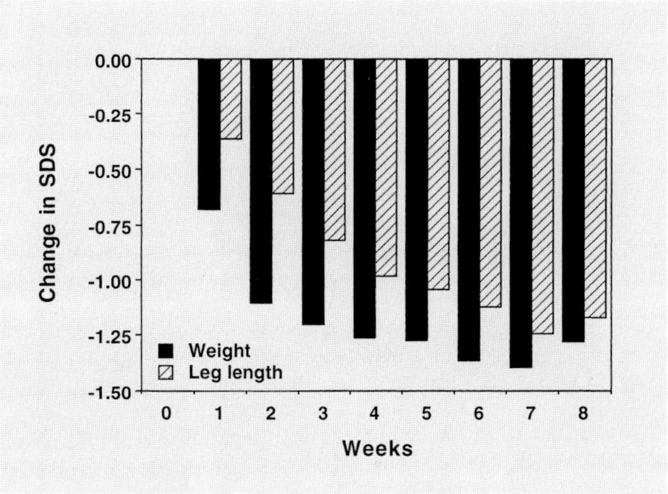

Fig. 4. Magnitude of the change of SDS (birth SDS subtracted from measurement SDS) in 122 pre-term infants in the first 8 post-natal weeks.

20]. A number of studies have shown that catch-up growth in SGA infants normally occurs within the first 3–6 months after birth but growth velocity is similar to non-SGA infants thereafter [21, 22]. Although the majority of catch-up growth occurs in the early post-natal period, there is some evidence to suggest that further catch-up growth may occur later. In a group of 3,650 infants 13.4% of SGA infants were below 2 standard deviation scores (SDS) in height at 1 year of age, whereas only 7.9% were at 18 years of age [23]. This study also confirmed the importance of length measurements. Twenty-two per cent of children who were short at birth remained short at 18 years as, compared with only 14% when birthweight was used to define SGA. Multiple regression analyses have shown that both birth length and mid-parental height are significantly related to the magnitude of catch-up growth [20, 23]. Although evidence suggests that predicted stature, based on mid-parental height, must be taken into account in the diagnosis and management of SGA, we are not aware of any studies that have compared outcome in SGA infants on the basis of this consideration.

Pre-Term Infants

Infants born prematurely may experience severe growth retardation of rapid onset. We have observed impaired weight and leg length gain – as indicated by a decrease in SDS from that at birth – in more than 90% of a group of 122 infants born at 33 weeks of gestation or below. Impaired growth was evident within 1 week of birth in the majority of infants and sustained over the first 8 post-natal weeks (fig. 3). Reduction in the rate of weight gain was greatest during the first 2–3 weeks after birth, while rate of leg length gain steadily decreased over the first 7 weeks (fig. 4). Average change in SDS was just over –1.25 for weight and –1.0 for length, but changes of more than –4 SDS for weight and –3 SDS for length were observed.

To assess the relationship between early growth impairment and subsequent growth pattern, we have made yearly measurements in 200 infants born at 35 weeks of gestation or below and 50 randomly selected term infants. Data collection at 60 months is nearly complete, and changes in SDS from birth measurements have been calculated for weight, length and head circumference (fig. 5–7). Early catch-up growth was seen in the majority of infants born at a gestational age of 29 weeks or above, and growth rate was similar to term infants from 12 months onwards. More immature infants demonstrated continued growth impairment for up to 24 months before catch-up was seen. Reduction in SDS for head circumference was more sustained than for weight or length. The proportions of infants with a reduction of between 1 and 2 and 2 or more SDS have been calculated at the different measurement intervals (fig. 8). At 60 months more than 40% of these infants have a head circumference that is at least 1 SDS lower than the birth measurement. Although skull

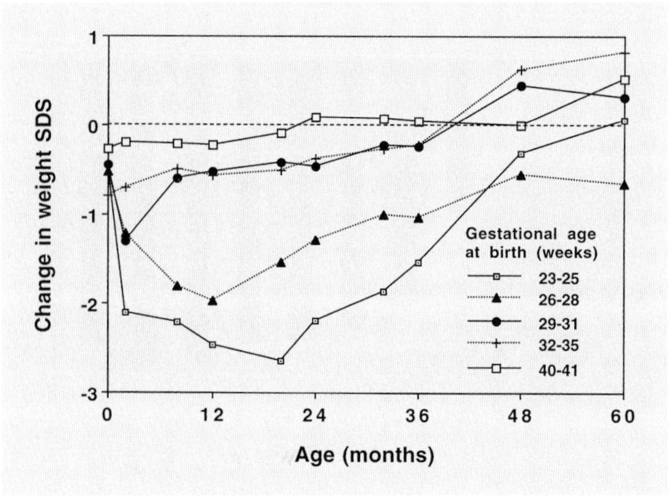

5

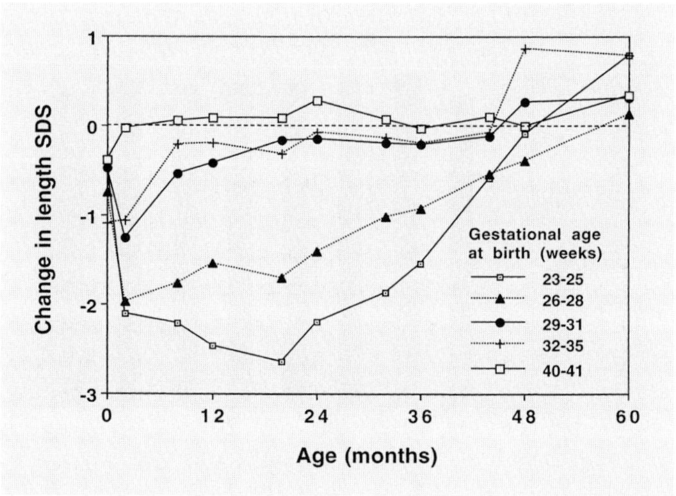

6

Fig. 5. Change in weight SDS in the first 5 post-natal years in a group of pre-term infants and 50 term infants, subdivided according to gestational age at birth.

Fig. 6. Change in length SDS in the first 5 post-natal years in a group of pre-term infants and 50 term infants, subdivided according to gestational age at birth.

Fig. 7. Change in occipitofrontal circumference (OFC) SDS in the first 5 post-natal years in a group of pre-term infants and 50 term infants, subdivided according to gestational age at birth.

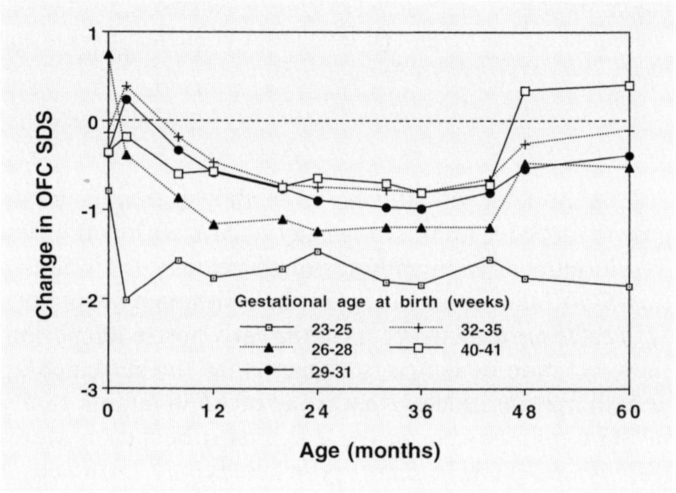

7

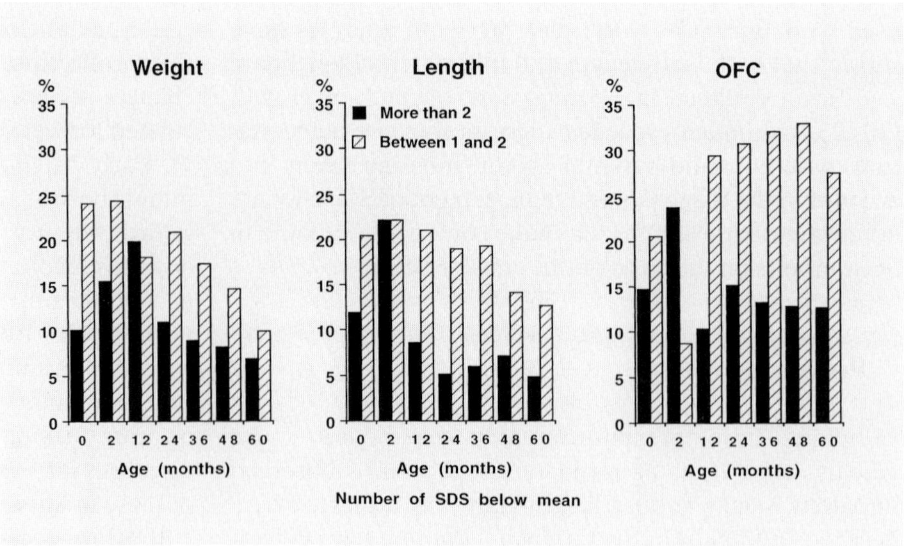

Fig. 8. Proportion of infants with a weight, length or occipitofrontal circumference (OFC) SDS ≤2 or >2 but ≤1 SDS below the population mean in a group of pre-term and term infants followed from birth until 60 months of corrected age.

moulding may explain some of these changes, a genuine reduction in head size may have occurred in a proportion of these infants.

Early Growth and Later Outcome

The pattern of early post-natal growth failure that we have demonstrated has been confirmed by a number of other studies. Gill et al. [24] evaluated early post-natal growth in 184 singleton survivors born between 23 and 29 weeks of gestation. Early post-natal growth patterns were very similar to those that we have demonstrated with growth failure being more marked and more prolonged in the most immature infants. From the 4th post-natal week, all subgroups demonstrated a more rapid rate of weight gain than would have been predicted from the intra-uterine growth rate, but mean weight remained below the 10th centile at 2 years of age. A multicentre study of longitudinal growth in 608 infants found that weight, height and head circumference at term remained significantly lower in low-birthweight pre-term infants than in those with a higher birthweight [25]. Growth patterns were lower than published standards for term infants of the same age and sex, and the patterns of growth differed by birthweight group. No catch-up growth was demonstrated by 12 months corrected age. Kitchen et al. [26] measured a cohort of 175 very-low-birthweight infants and found that mean incremental weight and height gain between 2 and 5 years was significantly lower than predicted and least in those who were below the 10th centile at 2 years. The majority of infants who were below the 10th centile for weight and height at 2 years remained so at 5 years [26]. Slightly different conclusions were reached in a study in which a cohort of 565 infants were measured at the same time intervals. Although stature at 2 years of age predicted stature at 5 years in a large proportion of children, 37% of those below the 10th centile at 2 years were above it by 5 years [27].

Studies in which data have been collected beyond 5 years are very limited. In a cohort of 171 infants with a birthweight of 2,000 g or less followed to 7 years of age, height remained significantly lower than predicted from mid-parental height and lower than in a cohort of normal birthweight peers [28]. Hack et al. [29] demonstrated catch-up growth in a large proportion of a group of 249 very-low-birthweight infants followed until 8 years of corrected age. While 54 and 60% were subnormal (below the 3rd centile) for weight and height, respectively, at term, these proportions had fallen to 8% by 8 years. The authors suggested that further significant catch-up growth was likely to occur during adolescence [29]. Powls et al. [30] assessed 137 very-low-birthweight children (in 70% of whom bone age was also measured) and 160 controls at 11–13.5 years of age. The low-birthweight infants had lower height, weight and head circumference than the controls, and short stature was not accounted for by differences in parental height or degree of pubertal development. Bone age was significantly advanced in the low-birthweight children. Using these measurements a final adult height below the 3rd centile was predicted for 17% and below the 10th centile for 33%. Weight was appropriate for height but head circumference appeared disproportionately low in the low-birthweight group [30].

Our observations and those of others have shown that early post-natal growth impairment is common, that poor growth may be sustained for many years and that catch-up growth may be incomplete. This may have significance beyond reduction in stature, as many phases of organ development and differentiation occur during this period and follow a defined chronological sequence. Lung development passes through several distinct phases during the latter part of pregnancy and early post-natal life, as the solid parenchyma becomes canalized and sacculated before alveolization occurs. The brain undergoes phases of myelination and dendrite and synapse formation which also follow a closely programmed timescale. If growth is impaired during these periods, it is conceivable that growth and differentiation of body organs may also be significantly affected. Very few studies have examined the relationship between growth failure and neurological status, and we are unaware of any that have assessed lung function. Ross et al. [31] assessed growth and neurodevelopmental status in a cohort of 86 premature infants at 1, 3 and 8 years. Infants with a poor neurodevelopmental score at 1 year were lighter, shorter and had smaller head circumferences than those with a developmental score within the normal range. At 3 years, the neurological status was significantly correlated with height and head circumference, while at 8 years only head circumference was related to neurological impairment [32, 33]. Astbury et al. [34] followed 235 very-low-birthweight infants from birth until 2 years of corrected age. Gross motor development was significantly delayed in those infants with weight below the 10th centile in comparison with those above. Major disability was more common in these infants (28 vs. 13%), and although only 29% of the cohort had a weight below the 10th centile, these infants represented 75% of those who were blind, 50% of those with serious motor delay and 42% of those with cerebral palsy [34]. No significant difference in the incidence or severity of other risk factors accounted for the apparent differences in neurodevelopmental outcome. In a follow-up of very-low-birthweight infants at 11–13.5 years of age, no relationship between weight or length and neurological outcome was demonstrable, but small head size was associated with lower IQ and mathematics and reading scores [30]. These limited data, therefore, suggest that there may be an association between growth impairment and subsequent neurodevelopmental outcome, though the evidence is far from conclusive. More rigorous and detailed assessment is required before this question can be fully answered.

Poor post-natal growth could have further implications, if the association between early growth and subsequent health proposed by the Barker hypothesis [35] is correct. A number of studies have shown a significant correlation between birthweight and subsequent risk of cardiovascular disease [36, 37], and an association between birthweight and cognitive function in adult life has also been reported [38]. There are some data to suggest that post-natal weight gain may also be relevant [39], while other studies have failed to show any relationship between adult morbidity and either birthweight or early growth [40]. It has been suggested that much of the evidence used to support the concept of fetal programming is flawed and may detract from the far more important contribution of post-natal growth [41].

Conclusions

There is no question that post-natal growth in pre-term infants may be severely impaired. There is substantial evidence to suggest that this impairment may be long lasting and a degree of permanent impairment may occur. There are undoubtedly concerns as to the impact that this growth impairment may have on other aspects of development, though evidence to confirm this is scarce. The lack of evidence should not lead to complacency, however. Further work is essential, so that the consequences of this growth impairment may be clearly defined, the aetiology explored and appropriate treatments investigated.

Acknowledgements

The authors wish to thank Joy Buckley for secretarial support and Serono Laboratories Ltd. (UK) for financial support of the long-term growth study.

References

1 Tanner JM: Fetus into Man: Physical Growth from Conception to Maturity. Ware, Castlemead, 1989.
2 Cooney K, Pathak U, Watson A: Infant's growth charts. Arch Dis Child 1994;71:159–160.
3 Child Growth Foundation: Growth Chart, Birth – 20 Years. London, Harlow, 1996.
4 Tanner JM, Whitehouse RH: Height and weight charts from birth to 5 years allowing for length of gestation: For use in infant welfare clinics. Arch Dis Child 1973;48:786–789.
5 Gairdner and Pearson Growth Chart. Welwyn Garden City, Castlemead, 1988.
6 Keen DV, Pearse RG: Weight, length and head circumference curve for boys and girls of between 20 and 42 weeks' gestation. Arch Dis Child 1988;63:1170–1172.
7 Smith DW, Truog W, Rogers JE, Greitzer LJ, Skinner AL, McCann JJ, Harvey MAS: Shifting linear growth during infancy: Illustration of genetic factors in growth from fetal life through infancy. J Pediatr 1976;89:225–230.
8 Greco L, Capasso A, De Fusco C, Paludetto R: Pulsatile weight increases in very low birthweight babies appropriate for gestational age. Arch Dis Child 1990;65:373–376.
9 Kaempf DE, Pfluger MS, Thiele AM, Hermanussen M, Linderkamp O: Influence of nutrition on growth in premature infants: Assessment by knemometry. Ann Hum Biol 1998;25:127–136.
10 Wales JKH, Carney S, Gibson AT: The measurement of neonates. Horm Res 1997;48 (suppl 1):2–10.
11 Altman DG, Hytten FE: Intra-uterine growth retardation: Lets be clear about it. Br J Obstet Gynaecol 1989;96:1127–1132.
12 Fancourt R, Campbell S, Harvey D, Norman AP: Follow-up study of small-for-dates babies. Br Med J 1976;i:1435–1437.
13 Gardosi J, Chang A, Kalyan B, Sahota D, Symonds EM: Customised antenatal growth charts. Lancet 1992;339:283–287.
14 Sanderson DA, Wilcox MA, Johnson IR: The individualised birthweight ratio: A new method of identifying intra-uterine growth retardation. Br J Obstet Gynaecol 1994;101:310–314.
15 Starfield B, Shapiro S, McCormick M, Bross D: Mortality and morbidity in infants with intrauterine growth retardation. J Pediatr 1982;101:978–983.
16 Goldenberg RL, DuBard MB, Cliver SP, Nelson KG, Blankson K, Ramey SL, Herman A: Pregnancy outcome and intelligence at age five years. Am J Obstet Gynecol 1996;175:1511–1515.
17 Sung IK, Vohr B, Oh W: Growth and neurodevelopmental outcome of very low birthweight infants with intrauterine growth retardation: Comparison with control subjects matched by birthweight and gestational age. J Pediatr 1993;123:618–624.
18 Low JA, Handley-Derry MH, Burke SO, Peters RD, Pater EA, Killen HL, Derrick EJ: Association of intrauterine fetal growth retardation and learning deficits at age 9 to 11 years. Am J Obstet Gynecol 1992;167:1499–1505.
19 Villar J, Smeriglio V, Martorell R, Brown CH, Klein RE: Heterogeneous growth and mental development of intrauterine growth-retarded infants during the first 3 years of life. Pediatrics 1984;74:783–791.
20 Albertsson-Wikland K, Karlberg J: Natural growth in children born small for gestational age with and without catch-up growth. Acta Paediatr Suppl 1994;399:64–70.
21 Tenovuo A, Kero P, Piekkala P, Korvenranta H, Sillanpaa M, Erkkola R: Growth of 519 small for gestational age infants during the first two years of life. Acta Paediatr Scand 1987;76:636–646.
22 Fitzhardinge PM, Inwood S: Long-term growth in small-for-date children. Acta Paediatr Scand Suppl 1989;349:27–33.

23 Karlberg J, Albertsson-Wikland K: Growth in full-term small-for-gestational-age infants: From birth to final height. Pediatr Res 1995; 38:733–739.

24 Gill A, Yu VYH, Bajuk B, Astbury J: Postnatal growth in infants born before 30 weeks' gestation. Arch Dis Child 1986;61:549–553.

25 Casey PH, Kraemer HC, Bernbaum J, Tyson JE, Sells JC, Yogman MW, Bauer CR: Growth patterns of low birthweight preterm infants: A longitudinal analysis of a large, varied sample. J Pediatr 1990;117:298–307.

26 Kitchen WH, Ford GW, Doyle LW: Growth and very low birth weight. Arch Dis Child 1989;64:379–382.

27 Qvigstad E, Verloove-Vanhorick SP, Ens-Dokkum MH, Schreuder AM, Veen S, Brand R, Oostdijk W, Ruys JH: Prediction of height achievement at 5 years of age in children born very preterm or with very low birth weight: Continuation of catch-up growth after two years of age. Acta Paediatr 1993;82:444–448.

28 Elliman A, Bryan E, Elliman A, Walker J, Harvey D: The growth of low-birth-weight children. Acta Paediatr 1992;81:311–314.

29 Hack M, Weissman B, Borawski-Clark E: Catch-up growth during childhood among very low-birth-weight children. Arch Pediatr Adolesc Med 1996;150:1122–1129.

30 Powls A, Botting N, Cooke RWI, Pilling D, Marlow N: Growth impairment in very low birthweight children at 12 years: Correlation with perinatal and outcome variables. Arch Dis Child 1996;75:F152–F157.

31 Ross G, Krauss AN, Auld PAM: Growth achievement in low-birth-weight premature infants: Relationship to neurobehavioural outcome at 1 year. J Pediatr 1983;103:105–108.

32 Ross G, Lipper EG, Auld PAM: Physical growth and developmental outcome in very low birth weight premature infants at 3 years of age. J Pediatr 1985;107:284–286.

33 Ross G, Lipper EG, Auld PAM: Growth achievement of very low birth weight premature children at school age. J Pediatr 1990;117: 307–309.

34 Astbury J, Orgill AA, Bajuk B, Yu VYH: Sequelae of growth failure in appropriate for gestational age, very low-birthweight infants. Dev Med Child Neurol 1986;28:472–479.

35 Barker DJP: Mothers, babies, and disease in later life. London, BMJ Publishing, 1994.

36 Forrester TE, Wilks RJ, Bennett FI, Simeon D, Osmond C, Allen M, Chung AP, Scott P: Fetal growth and cardiovascular risk factors in Jamaican schoolchildren. BMJ 1996;312:156–160.

37 Rich-Edwards JW, Stampfer MJ, Manson JE, Rosmer B, Hankinson SE, Colditz GA, Willett WC, Hennekens CH: Birth weight and risk of cardiovascular disease in a cohort of women followed up since 1976. BMJ 1997;315:396–400.

38 Sorensen HT, Sabroe S, Olsen J, Rothman KJ, Gillman MW, Fischer P: Birth weight and cognitive function in young adult life: Historical cohort study. BMJ 1997;315:401–403.

39 Fall CHD, Vijayakumar M, Barker DJP, Osmond C, Duggleby S: Weight in infancy and prevalence of coronary heart disease in adult life. BMJ 1995;310:17–19.

40 Laor A, Stevenson DK, Shemer J, Gale R, Seidman DS: Size at birth, maternal nutritional status in pregnancy, and blood pressure at age 17: Population based analysis. BMJ 1997;315: 449–453.

41 Lucas A, Fewtrell MS, Cole TJ: Fetal origins of adult disease – the hypothesis revisited. BMJ 1999;319:245–249.

Hormone Research

Horm Res 2000;53(suppl 1):50–54

Growth Hormone and Growth Factors during Perinatal Life

Hartmut A. Wollmann

Section Pediatric Endocrinology, University Children's Hospital Tübingen, Germany

Key Words
Growth hormone · Insulin-like growth factor · Perinatal life · Fetus · Growth

Abstract
Pituitary growth hormone (GH) is present in early pregnancy in the fetal circulation. The concentrations are higher than ever found during life, due to an unrestrained, basal secretion. GH receptors develop around midpregnancy, when they are present in low concentrations, and there is a rapid increase during the first months of life. The function of fetal GH – characterized by a nearly complete GH resistance – is largely unclear: there is only a small effect on longitudinal growth, and the regulation of growth factors is independent of GH. Possibly, metabolic effects of GH on fat and glucose metabolism and body composition are of greater importance. During the first months of life, the rapid fetal (GH-independent, nutrition-dependent) growth decelerates, a process that is partly compensated by the onset of GH-dependent longitudinal growth.

Introduction

Compared with later life, fetal and perinatal longitudinal growth is an extremely rapid process. During midpregnancy, the fetus grows with a velocity of approximately 10 cm/month; around birth and during the first months of life, a growth velocity of 25 cm/year is normal. By comparison, the pubertal growth spurt is rather slow (fig. 1). During childhood, there is a central endocrine control of growth through pituitary growth hormone (GH) and the GH-dependent growth factors. In contrast, fetal and perinatal growth regulation is much more complex and far less understood.

Growth Hormone

In the fetal circulation, pituitary GH can be determined from the 12th week of gestation onwards, whereas expression in pituitary cells starts earlier. By mid-pregnancy, plasma levels are in the range of 100 ng/ml. These are the highest GH concentrations reached under physiological conditions throughout life. In late gestation, serum GH concentrations in the fetus decrease [1]. The GH serum pattern during gestation reflects an unrestrained secretion at mid-gestation and a progressively dominant inhibitory tone as a consequence of the maturation of the hypothalamic-pituitary function (fig. 2).

KARGER
Fax + 41 61 306 12 34
E-Mail karger@karger.ch
www.karger.com

0301–0163/00/0537–0050$17.50/0

Accessible online at:
www.karger.com/journals/hre

Dr. H.A. Wollmann
Section Pediatric Endocrinology, University Children's Hospital
Hoppe-Seyler-Strasse 1
D–72076 Tübingen (Germany)
Tel. +49 7071 2983781, Fax +49 7071 294157, E-Mail htwollma@med.uni-tuebingen.de

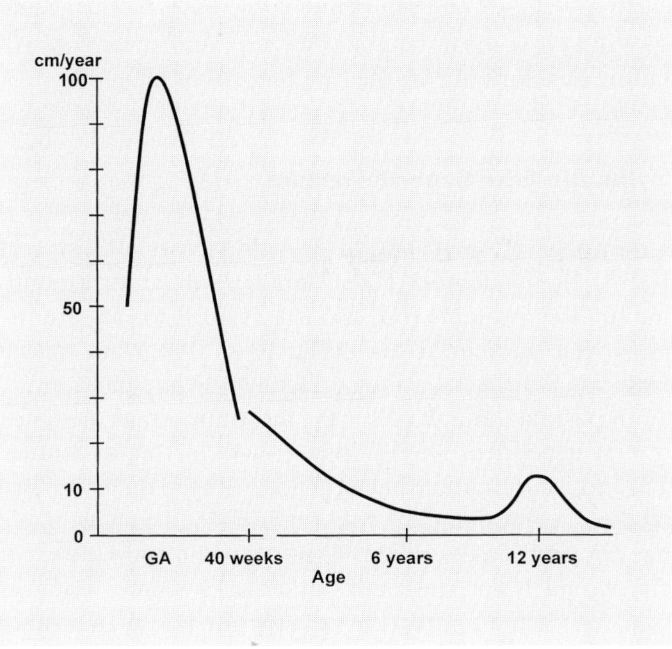

Fig. 1. Growth velocity from fetal life to puberty (qualitative). GA = Gestational age.

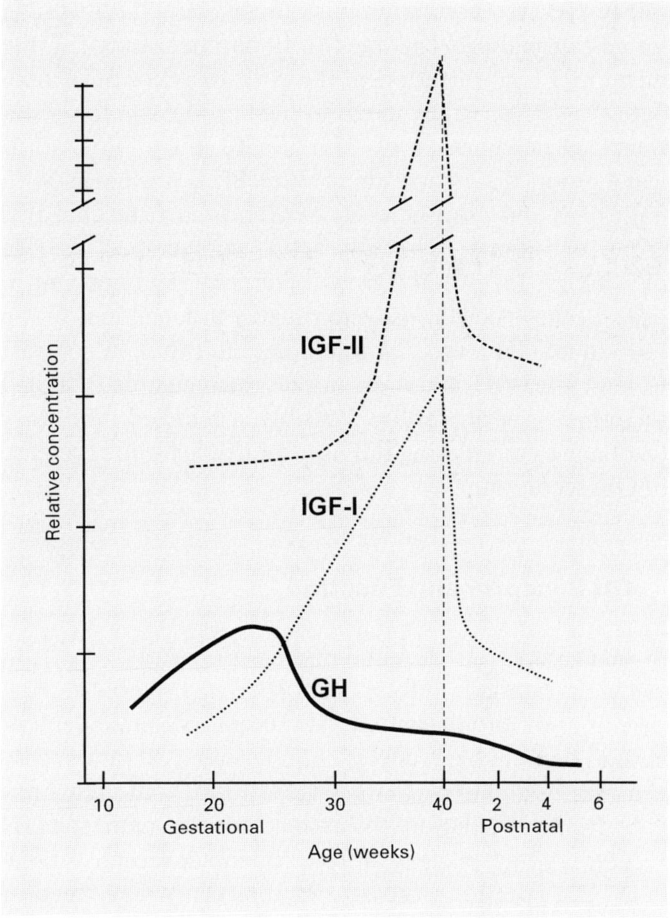

Fig. 2. Relative concentrations of GH and IGF-I and IGF-II during gestation and perinatally.

In premature infants, GH concentrations in the circulation are elevated, and the secretion follows a strictly pulsatile pattern [2]. In very low birthweight premature infants, serum GH concentrations remain elevated until around term, indicating that – in spite of extrauterine life – the fetal situation continues postnatally [3].

As is the case for other hypothalamic-pituitary hormones, control of GH secretion becomes progressively mature during the last half of gestation. At term [4], pituitary GH secretion is still enhanced compared with childhood, but there is a rapid decline of serum concentrations within the first postnatal days. Regulation by somatostatin and GH-releasing factor [5] is intact in term newborns. As the infant matures, there is a decrease in GH pulse frequency, pulse amplitude and nadir GH levels [4], resulting in a decrease of mean serum GH concentrations postnatally. During the first 3 months, wake and sleep values are not different. The typical childhood circadian pattern, with predominant GH secretion during the night, develops during the first 6 months [6]. There is a parallel development with other hormonal regulatory circuits, e.g., adrenal cortisol secretion, which also develops to the mature circadian secretory pattern with high concentrations in the morning and low values in the late evening during the first 6 months postnatally.

GH and Fetal Growth

Although GH is abundant in the fetal circulation, at least in the second half of gestation, it is not a major hormonal regulator of fetal growth, as anencephalic fetuses, which are not able to produce significant amounts of pituitary GH, are characterized by a nearly normal weight and length. The same is true for otherwise healthy infants suffering from severe congenital GH deficiency due to a gene defect or a morphologic defect of the hypothalamic-pituitary system: with a mean of 49 cm, their length at birth is nearly normal [7], but these infants are relatively obese. This could indicate that in perinatal – as in adult – life GH affects metabolism and body composition rather than longitudinal growth.

In term and preterm infants born 'small for gestational age' (SGA), GH concentrations in cord blood were significantly elevated compared with those being 'appropriate for gestational age' or large [8]. In another study [9], the mean GH concentration in the cord blood of SGA neonates was 72 ng/ml, whereas in the appropriate for gestational age newborns it was 37 ng/ml. In spite of this significant group difference, there were SGA newborns with both high and normal levels of GH, indicating that this group of infants is heterogeneous with respect to GH physiology. Insulin-like growth factor (IGF) I concentrations in these newborns were roughly independent of the individual GH levels. This demonstrates that, in contrast to later life, GH is not an important regulator of IGF-I secretion in perinatal life. Possibly, a reduced inhibitory feedback by circulating IGF is responsible for the increase of GH secretion.

GH Receptor and Function

GH receptor mRNA and protein is present by the 15th week of gestation on fetal chondrocytes, osteoblasts, epidermis and fibroblasts [10]. GH receptors can be demonstrated on fetal hepatocytes at about week 30 in low concentrations and in placental tissues throughout pregnancy [11]. In the circulation, the extracellular domain, the GH receptor, which is cleaved from the intact receptor molecule by a specific protease and represents GH receptor expression, can be measured as GH-binding protein (GHBP) from midpregnancy in low concentrations. In premature infants, low concentrations of GHBP [3] are measured which start to increase after term. The functional relevance of the GH receptor, at least for fetal growth and development, is unclear. The lack of a significant number of GH receptors explains the physiologic 'GH resistance' (with respect to longitudinal growth and IGF-I regulation) in prenatal and early postnatal life. Accordingly, the secretion of IGF-I, which is mediated by activation of the GH receptor, is mostly independent of GH in pre- and early postnatal life. Probably, nutrient supply is the most important regulator of IGF-I during this time, whereas GH additionally starts to act on IGF-I release during infancy.

Postnatally, GH receptor expression is gradually upregulated in many tissues [12]. In the circulation, this effect is measurable by the increase of GHBP concentrations [13]. In parallel with the upregulation of GH receptor expression, there is an increasing functional role of GH with respect to longitudinal growth. This effect can be demonstrated clinically by the growth pattern of infants with severe, congenital GH deficiency: length at birth is slightly reduced (standard deviation score – 1.4, corresponding to a mean length of 49 cm), and severe growth failure develops during the first months of life [14].

Insulin-Like Growth Factors

Among the polypeptide growth factors, IGF-I and IGF-II, their receptors (types I and II) and the corresponding binding proteins (IGFBP-1 to IGFBP-6) [15] are the most well studied factors that play an important role in embryonal differentiation and fetal growth regulation.

In postnatal life, IGF-I is the most important mediator of GH-dependent longitudinal growth, acting as a mitogen and a differentiation factor according to many in vitro studies. A physiological role for IGF-II in postnatal life has not yet been described. The liver is the main source of circulating IGFs, but many other cells secrete IGFs as well, indicating a paracrine or autocrine effect. In postnatal life, IGFBP-3 is the most abundant binding protein in the circulation, binding more than 90% of the IGFs. Its secretion – as that of the IGFs – is mostly dependent on GH secretion, nutrition and liver function.

In the human fetus, IGF-I is detectable in several tissues from 9 weeks of pregnancy onwards [16], whereas expresssion at the cellular level starts earlier. By midpregnancy, IGFs are abundant in many fetal tissues, such as lung, kidney, gut, muscle and skin [17], with the highest concentrations being in the lung. In the serum of midpregnancy fetuses, levels of IGF-II exceed IGF-I by about sevenfold, and there is a further parallel increase of both IGFs during the last trimester to values that are two to three times higher at term (fig. 2).

Around midpregnancy, receptors for IGF-I and IGF-II are present in nearly all fetal tissues [18]. IGFBPs can be detected in fetal serum, with a pattern resembling that of hypopituitarism in postnatal life: elevated concentrations of IGFBP-2 and relatively low concentrations of IGFBP-3 [19, 20].

At birth, there is a direct positive relationship between IGF-I levels in umbilical cord serum and birthweight [21–23], suggesting a functional role of at least IGF-I in fetal growth. In children with intrauterine growth retardation, IGF concentrations are significantly reduced compared with normals [24, 25].

Experiments with knockout animals – which cannot be discussed here in detail – give detailed information about the mechanism of action of IGFs [26, 27] and demonstrate convincingly the role of IGF-I and IGF-II in fetal growth.

Direct evidence for an influence of IGF-I on human fetal growth came from a patient with a homozygous defect of the IGF-I gene [28]. This patient with severe growth failure before and after birth, sensorineural deafness, dysmorphic features, microcephaly and mental retardation had two exons of his IGF-I gene missing. This illustrates the complex influence of IGF-I, not only on fetal longitudinal growth but also on central nervous system development.

The regulation of fetal IGF expression is complex and only partially understood. Whereas in postnatal life GH is the most important regulator of IGF synthesis, other factors are of greater importance prenatally. The most important factor regulating IGFs in late gestation is probably adequate nutrition.

According to the results of Gluckman [29], fetal growth during the second half of gestation and size at birth are not primarily determined by genetic factors, but by the uterine environment which normally constrains fetal growth. This constraint is largely mediated by fetal substrate supply caused by maternal disease, malnutrition or by placental malfunction. Physiologically, insulin as the mediator of this nutritional effect directly influences circulating IGF-I and IGF-II concentrations in the fetus [30]. Fetal hyperinsulinism due to a permanent hyperglycemia of the mother (badly controlled diabetes) induces an increase of IGF-I expression which is measurable in cord blood of these macrosomic newborns [31]. In insulin deficiency, the fetus exhibits growth retardation, muscle wastage and reduced body fat as well as significantly reduced IGF-I concentrations. As insulin is the most important anabolic hormone [32], on the other hand, it has direct growth-promoting effects in the fetus [33], and it is difficult to differentiate between the effects of these two synergistic factors. At the receptor level, the interaction of insulin and IGF-I is even more complex, as insulin and IGF-I receptors share homology, and both ligands can bind to both receptors.

Whether IGFBPs have influence on fetal growth is widely unknown. IGFBP-3, the predominant binding protein in childhood and adolescence, is measurable in low concentrations from midpregnancy onwards, without obvious functional significance. Possibly, IGFBP-1, which has a short half-life and is inversely regulated by insulin, is involved in the regulation of fetal growth [22]. It antagonizes the biological effect of free IGF-I and modulates IGF-I bioactivity. In the human neonate at term, IGFBP-1 is one of the major binding proteins of IGF-I. In contrast to IGF-1, the concentrations of IGFBP-1 are significantly elevated in SGA newborns as well as in cases of fetal hypoxia [22, 34]. The highest levels of IGFBP-1 are found in those SGA cases in whom uteroplacental insufficiency can be demonstrated by Doppler investigation.

Possibly, macrosomic babies of pregnant mothers with poorly controlled diabetes mellitus [30] are a good model illustrating the modulating effect of IGFBP-1 on fetal growth. Insulin is an important regulator of IGFBP-1, and there is an inverse relationship between the two: hyperinsulinemia in the fetus secondary to hyperglycemia of the mother inhibits IGFBP-1 production in the fetal liver. A reduced concentration of the binding protein enhances the levels of free IGF-I, leading to fetal overgrowth [35].

Conclusions

Fetal growth is determined by several factors: genetic constitution, nutritional status and pathological conditions (e.g., maternal diabetes or uterine constraints). In spite of extensive investigation, a distinct and specific endocrine mechanism essential for fetal growth has not yet been described. IGF-I and IGF-II play a central role in the regulation of fetal growth by autocrine or paracrine mechanisms. Their regulation depends on different factors – insulin and nutrition being most important – and is different from postnatal regulation. Until term, growth factors and longitudinal growth are independent of GH. However, GH-dependent growth, which progressively starts during the 1st year of life, contributes less than 50% to total postnatal growth. As an example, a woman with a severe GH deficiency reaches a final height of approximately 120 cm, corresponding to about 70 cm of GH-independent, postnatal growth. Therefore, the mechanisms controlling prenatal growth probably are still active in postnatal life and quantitatively even more important than direct GH-dependent growth.

References

1 Gluckman PD, Grumbach MM, Kaplan SL: The neuroendocrine regulation and function of growth hormone and prolactin in the mammalian fetus. Endocr Rev 1981;2:363–395.

2 Miller JD, Wright NM, Esparza A, Jansons R, Yang HC, Hahn H, Mosier HD Jr: Spontaneous pulsatile growth hormone release in male and female premature infants. J Clin Endocrinol Metab 1992;75:1508–1513.

3 Wollmann HA: Insulin-ähnliche Wachstumsfaktoren und ihre Bindungsproteine bei frühgeborenen Kindern: Verlauf, Regulation und physiologische Bedeutung. Habilitationsschrift. Eberhard-Karls-Universität, Tübingen, 1999.

4 Miller JD, Esparza A, Wright NM, Garimella V, Lai J, Lester SE, Mosier HD Jr: Spontaneous growth hormone release in term infants: Changes during the first four days of life. J Clin Endocrinol Metab 1993;76:1058–1062.

5 Shimano S, Suzuki S, Nagashima K, Yagi H, Sakaguchi M, Kuroume T: Growth hormone responses to growth hormone releasing factor in neonates. Biol Neonate 1985;47:367–370.

6 Vigneri R, D'Agata R: Growth hormone release during the first year of life in relation to sleep-wake periods. J Clin Endocrinol Metab 1971;33:561–563.

7 Gluckman PD: Growth hormone deficiency diagnosed and treated in the first 2 years of life: Evidence of the role of growth hormone in human perinatal growth; in Ranke MB, Gunnarsson R (eds): Progress in Growth Hormone Therapy. Mannheim, J & J Verlag, 1994, pp 88–96.

8 de Zegher F, Kimpen J, Raus J, Vanderschueren Lodeweyckx M: Hypersomatotropism in the dysmature infant at term and preterm birth. Biol Neonate 1990;58:188–191.

9 Varvarigou A, Vagenakis AG, Makri M, Beratis NG: Growth hormone, insulin-like growth factor-I and prolactin in small for gestational age neonates. Biol Neonate 1994;65:94–102.

10 Werther GA, Haynes K, Waters MJ: Growth hormone (GH) receptors are expressed on human fetal mesenchymal tissues – identification of messenger ribonucleic acid and GH-binding protein. J Clin Endocrinol Metab 1993;76:1638–1646.

11 Hill DJ, Riley SC, Bassett NS, Waters MJ: Localization of the growth hormone receptor, identified by immunocytochemistry, in second trimester human fetal tissues and in placenta throughout gestation. J Clin Endocrinol Metab 1992;75:646–650.

12 Massa G, de Zegher F, Vanderschueren Lodeweyckx M: Serum growth hormone-binding proteins in the human fetus and infant. Pediatr Res 1992;32:69–72.

13 Silbergeld A, Lazar L, Erster B, Keret R, Tepper R, Laron Z: Serum growth hormone binding protein activity in healthy neonates, children and young adults: Correlation with age, height and weight. Clin Endocrinol (Oxf) 1989;31:295–303.

14 Gluckman PD, Gunn AJ, Wray A, Cutfield WS, Chatelain PG, Guilbaud O, Ambler GR, Wilton P, Albertsson Wikland K: Congenital idiopathic growth hormone deficiency associated with prenatal and early postnatal growth failure. The International Board of the Kabi Pharmacia International Growth Study. J Pediatr 1992;121:920–923.

15 Sara VR, Hall K: Insulin-like growth factors and their binding proteins. Physiol Rev 1990;70:591–614.

16 Han VK, Lund PK, Lee DC, D'Ercole AJ: Expression of somatomedin/insulin-like growth factor messenger ribonucleic acids in the human fetus: Identification, characterization, and tissue distribution. J Clin Endocrinol Metab 1988;66:422–429.

17 D'Ercole AJ, Hill DJ, Strain AJ, Underwood LE: Tissue and plasma somatomedin-C/insulin-like growth factor I concentrations in the human fetus during the first half of gestation. Pediatr Res 1986;20:253–255.

18 Funk B, Kessler U, Eisenmenger W, Hansmann A, Kolb HJ, Kiess W: Expression of the insulin-like growth factor-II/mannose-6-phosphate receptor in multiple human tissues during fetal life and early infancy. J Clin Endocrinol Metab 1992;75:424–431.

19 Chard T: Insulin-like growth factors and their binding proteins in normal and abnormal human fetal growth. Growth Regul 1994;4:91–100.

20 Lassarre C, Hardouin S, Daffos F, Forestier F, Frankenne F, Binoux M: Serum insulin-like growth factors and insulin-like growth factor binding proteins in the human fetus. Relationships with growth in normal subjects and in subjects with intrauterine growth retardation. Pediatr Res 1991;29:219–225.

21 Verhaeghe J, Van Bree R, Van Herck E, Laureys J, Bouillon R, Van Assche FA: C-peptide, insulin-like growth factors I and II, and insulin-like growth factor binding protein-1 in umbilical cord serum: Correlations with birth weight. Am J Obstet Gynecol 1993;169:89–97.

22 Wang HS, Chard T: The role of insulin-like growth factor-I and insulin-like growth factor-binding protein-1 in the control of human fetal growth. J Endocrinol 1992;132:11–19.

23 Fall CH, Pandit AN, Law CM, Yajnik CS, Clark PM, Breier B, Osmond C, Shiell AW, Gluckman PD, Barker DJ: Size at birth and plasma insulin-like growth factor-1 concentrations. Arch Dis Child 1995;73:287–293.

24 Mirlesse V, Frankenne F, Alsat E, Poncelet M, Hennen G, Evain Brion D: Placental growth hormone levels in normal pregnancy and in pregnancies with intrauterine growth retardation. Pediatr Res 1993;34:439–442.

25 Wollmann HA: Intrauterine growth retardation; in Kelnar CJH, Stirling HF, Saenger P, Savage MO (eds): Growth Disorders – Pathophysiology and Treatment. London, Chapman & Hall, 1998, pp 661–672.

26 Baker J, Liu JP, Robertson EJ, Efstratiadis A: Role of insulin-like growth factors in embryonic and postnatal growth. Cell 1993;75:73–82.

27 Liu JP, Baker J, Perkins AS, Robertson EJ, Efstratiadis A: Mice carrying null mutations of the genes encoding insulin-like growth factor I (Igf-1) and type 1 IGF receptor (Igf1r). Cell 1993;75:59–72.

28 Woods KA, Camacho Hubner C, Savage MO, Clark AJ: Intrauterine growth retardation and postnatal growth failure associated with deletion of the insulin-like growth factor I gene. N Engl J Med 1996;335:1363–1367.

29 Gluckman PD: Clinical review 68: The endocrine regulation of fetal growth in late gestation: The role of insulin-like growth factors. J Clin Endocrinol Metab 1995;80:1047–1050.

30 Fowden AL: The role of insulin in prenatal growth. J Dev Physiol 1989;12:173–182.

31 Roth S, Abernathy MP, Lee WH, Pratt L, Denne S, Golichowski A, Pescovitz OH: Insulin-like growth factors I and II peptide and messenger RNA levels in macrosomic infants of diabetic pregnancies. J Soc Gynecol Invest 1996;3:78–84.

32 Fowden AL: Insulin deficiency: Effects on fetal growth and development. J Paediatr Child Health 1993;29:6–11.

33 Susa JB, McCormick KL, Widness JA, Singer DB, Oh W, Adamsons K, Schwartz R: Chronic hyperinsulinemia in the fetal rhesus monkey: Effects on fetal growth and composition. Diabetes 1979;28:1058–1063.

34 Wang HS, Lim J, English J, Irvine L, Chard T: The concentration of insulin-like growth factor-I and insulin-like growth factor-binding protein-1 in human umbilical cord serum at delivery: Relation to fetal weight. J Endocrinol 1991;129:459–464.

35 Iwashita M, Kobayashi M, Matsuo A, Nakayama S, Mimuro T, Takeda Y, Sakamoto S: Feto-maternal interaction of IGF-I and its binding proteins in fetal growth. Early Hum Dev 1992;29:187–191.

Hormone Research

Horm Res 2000;53(suppl 1):55–59

Long-Term Results with Growth Hormone Therapy in Idiopathic Hypopituitarism

Sergio Bernasconi[a] Teresa Arrigo[b] Malgorzata Wasniewsk[b]
Lucia Ghizzoni[c] Caterina Ruggeri[b] Giuseppe Di Pasquale[b]
Alessandra Vottero[c] Filippo De Luca[b]

Departments of Paediatrics, Universities of [a]Modena and Reggio Emilia, [b]Messina and [c]Parma, Italy

Key Words
Growth hormone deficiency · Replacement therapy · Idiopathic hypopituitarism

Abstract
More than 30 years after its introduction, growth hormone (GH) treatment is well established in children with GH deficiency. Nevertheless, the long-term results of this therapy, expressed as height, are generally considered unsatisfactory. We report on results obtained in a group of GH-deficient children who were treated with daily injections of recombinant GH within the first 5 years of life and who reached an adult height very close to their target height. The full catch-up growth to the target height demonstrated in these patients suggests that replacement therapy should be started early and continued until adulthood. Height at onset of puberty is an important variable which might significantly influence the adult height. The significant and prolonged influence of birth weight on growth response to GH therapy underlines the important role of fetal growth in planning early treatment of GH-deficient children.

Adult Height of Growth Hormone Deficient Patients

It is well established that children with growth hormone (GH) deficiency (GHD) who are left untreated reach an adult height (AH) that is lower than normal, ranging from −3.1 to −7.4 standard deviation scores (SDS) [1–3]. Since 1958, pituitary-extracted human GH (hGH) has been available to improve the growth rate of patients with this disease [4], but several studies have shown that, although the final stature could be improved, a large majority of these patients do not achieve a normal AH [5, 6]. These unsatisfactory results might be ascribed to several reasons, including impurity of GH preparations, suboptimal dosage, low frequency of injections (two to three times a week) and short treatment regimens, mainly due to the difficulties of obtaining hGH supplies. Indeed, a positive correlation between GH dosage and growth velocity has been demonstrated [7, 8], and an increased growth rate has been shown after shifting from a three-times-a-week regimen to daily GH injections [8, 9].

On the basis of these findings and with the subsequent availability of unlimited supplies of recombinant GH

KARGER
Fax + 41 61 306 12 34
E-Mail karger@karger.ch
www.karger.com

0301–0163/00/0537–0055$17.50/0

Accessible online at:
www.karger.com/journals/hre

Prof. S. Bernasconi
Dipartimento di Scienze Ginecologiche, Ostetriche e Pediatriche
Università di Modena e Reggio Emilia, Policlinico, Largo del Pozzo 71
I-41100 Modena (Italy)
Tel. +39 059 424570, Fax +39 059 424583, E-Mail berna@unimo.it

(rGH), the majority of children with GHD have been treated regularly over the past 14 years with higher doses of GH, daily injections and for longer periods than when hGH was the only source of replacement therapy. Despite these improvements in therapy, however, published results in terms of AH continue to be unsatisfactory (table 1); in these studies, GHD of different aetiologies (idiopathic and organic) were analyzed together, the age at start of treatment was variable [10], and some patients did not receive daily therapy [10–12]. These factors might have contributed to the controversial results obtained [10].

Adult Height of Children with Idiopathic GHD Treated with Long-Term GH Therapy

In order to evaluate a homogeneous group of patients, the results of long-term GH therapy have been analyzed in patients with idiopathic GHD (table 2) [13–16]. Although some of the results were promising, the AH remained lower than normal, suggesting that therapeutic regimens could be further optimized. In fact, most patients did not follow the same treatment schedule, especially at the beginning of therapy. To overcome this possible bias, we have analyzed a more homogeneous group of patients (table 3). All patients were affected by idiopathic GHD and were treated with rGH six to seven times a week by subcutaneous injections since the beginning of the therapy. In the study by Cacciari et al. [17] all patients were diagnosed as having isolated GHD on the basis of 'classic' pharmacological tests and/or spontaneous nocturnal GH secretion. Moreover, all patients showed a normal pituitary on magnetic resonance imaging and underwent GH secretion re-evaluation at least once during puberty to exclude either hidden organic causes of hypopituitarism [18] or transient GHD [19]. Nevertheless, the mean AH of this group of patients was not different from that of the previous group (tables 1 and 2).

To date, the best results have been obtained in a Swedish study [20] with a mean AH of –0.32 SDS being achieved and in which patients received lower GH doses (0.65–0.67 IU/kg/week) than in the North American trial (1.0 IU/kg/week) [10, 21], suggesting that GH dosage might not be a limiting factor, at least not in the range used. Interestingly, 59% of the children enrolled in the KIGS study (Pharmacia & Upjohn International Growth Study) were diagnosed as having multiple pituitary hormone deficiency.

Grumbach et al. [22] showed no significant difference in AH between a group of children treated with hGH and another group treated with rGH; the former, however, were younger and shorter at the start of therapy than the rGH-treated group.

Table 1. Final height of patients with GHD after long-term treatment with rGH

Authors	Year of publication	Patients	Final height SDS
Blethen et al. [10]	1997	121	–0.7 ± 1.3 (male) –0.7 ± 1.1 (female)
Severi [11]	1995	71	–1.63 ± 0.14
Braemswig et al. [12]	1995	24[a]	–0.77 ± 0.88

[a] Patients with 'complete' GHD.

Table 2. Final height of idiopathic GHD children

Authors	Year of publication	Idiopathic GHD		Multiple pituitary hormone deficiency	
		n	final height SDS	n	final height SDS
Frisch and Birnbacher [13]	1995	5 (male) 5 (female)	–2.8 ± 1.3 –3.3 ± 0.7	14 (male) 9 (female)	–1.9 ± 0.9 –1.6 ± 1.2
Rikken et al. [14]	1995	25	–1.95 ± 1.30		
Birnbacher et al. [15]	1998	25 (male)	–2.30 ± 1.2[a] –1.2 ± 1.0[b]		
Cutfield et al. [16] (KIGS)	1999	369	–1.5 (–3.1 to +0.2)[c]		

[a] Spontaneous puberty.
[b] Induced puberty.
[c] Median (10th–90th percentile).

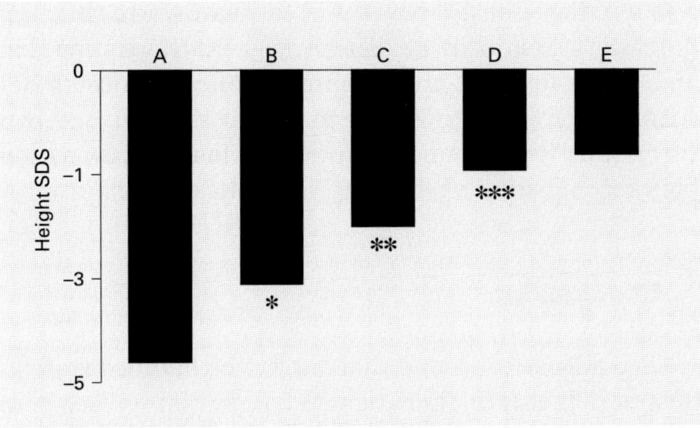

Fig. 1. Average values of height (SDS) during treatment in patients treated within the first 5 years of life. A = Before treatment; B = after 2 years of treatment; C = at onset of puberty; D = adult; E = target. * p < 0.0025 vs. A; ** p < 0.0005 vs. B; *** p < 0.0005 vs. C. [From ref. 30, with permission.]

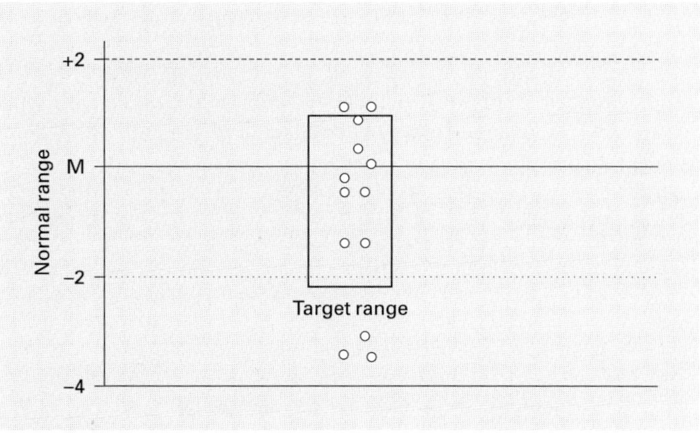

Fig. 2. Individual AH values (SDS) of early-treated GH-deficient patients. The dotted lines represent the normal limits; the box represents the target range. [From ref. 30, with permission.]

Table 3. Chronological age at start of GH treatment

Authors	Year of publication	Age at start of treatment, years (mean ± SD)
Blethen et al. [10]	1997	11.30±2.10
Braemswig et al. [12]	1995	12.95±2.13
Frisch and Birnbacher [13]	1995	9.90±2.80
Rikken et al. [14]	1995	8.49±3.21
Birnbacher et al. [15]	1998	8.90±3.20
Cutfield et al. [16] (KIGS)	1999	9.8 (4.6–12.9)[a]
Cacciari et al. [17]	1997	12.2±1.7

[a] Median (10th–90th percentile).

Influence of Early GH Treatment on AH in Patients with Idiopathic GHD

Among the factors positively or negatively related to AH, a lower chronological age at the start of therapy has been hypothesized as being one of the predictive factors of the final outcome [13, 14, 22–25]. Significant correlations between several pretreatment parameters (e.g., mid-parental height, height at start of therapy and birth weight) and final height were shown in a large number of patients affected by idiopathic GHD and who were enrolled in the KIGS [20].

In many studies, however, the age at the start of treatment ranged from 8.4 to 12.9 years [10, 12–17], suggesting the possibility of a delay in the diagnosis. This delay could also explain, at least in part, the suboptimal therapeutic results (table 3).

AH of Patients with GHD Treated with rGH Since the First Years of Life

In the past few years, there have been a few studies reporting long-term growth results after rGH therapy in early-treated GH-deficient patients [26–29], but only one report on AH has been published [30]. In this report, we were able to confirm that early GH replacement therapy produced a normal and target-related AH in a group of patients with idiopathic GHD (fig. 1) [30].

We have analyzed the AH of 13 children with idiopathic GHD who were treated with rGH. They presented with an average growth failure of −4.1 SDS. GH replacement therapy (0.6–0.7 IU/kg/week) started within the first 5 years of age in all patients and lasted at least 9 years, until adulthood [30].

In summary, our study demonstrated the following. (1) The average AH SDS reached by our patients was not significantly different from their target height SDS, 85% of the patients achieved an AH within both target range and normal limits (fig. 2); (2) the cumulative height gain under GH therapy (3.2 ± 1.9 SDS) was the result of both prepubertal (2.0 ± 2.1 SDS) and pubertal height gain (1.2 ± 1.8 SDS), the AH positively correlated with both variables; (3) although the treatment was started before

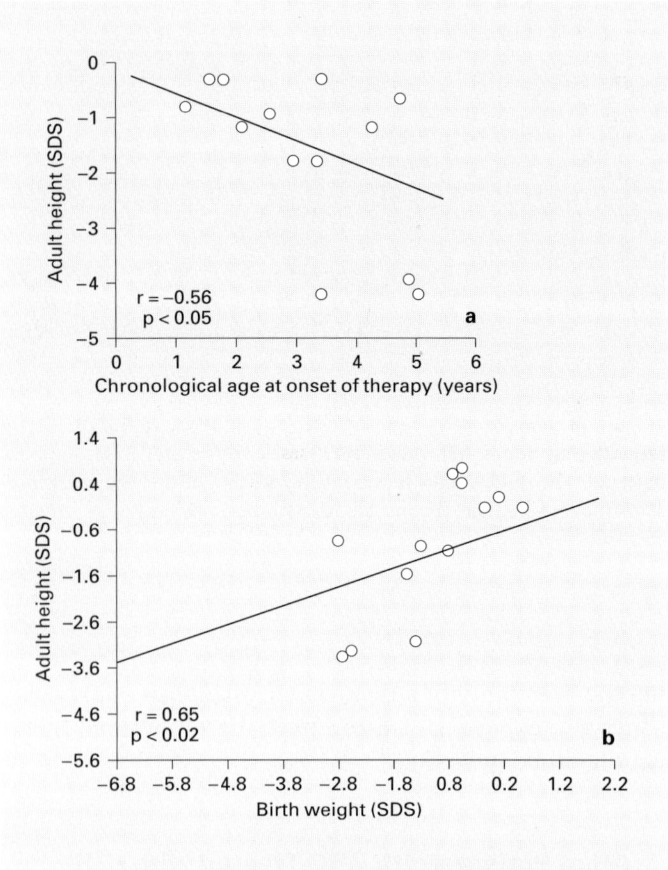

Fig. 3. Correlations between chronological age (years) at onset of GH therapy and AH (SDS) (**a**) and between birth weight (SDS in relation to the gestational age) and AH (SDS) (**b**) in early-treated GH-deficient patients.

5 years of age, the AH negatively correlated with the chronological age at start of therapy (fig. 3a), suggesting that an even earlier start might be worthwhile, and (4) the AH correlated positively with birth weight (fig. 3b), and this correlation was significantly higher than the correlation between AH and target height.

Conclusions

The full catch-up growth to target height that was demonstrated in our GHD patients suggests that replacement therapy should be started within the first 5 years of life and prolonged until adulthood. Height at onset of puberty is an important variable which might significantly influence the final height. The significant and prolonged influence of the birthweight on growth response to GH therapy underlines the important role of fetal growth in planning early treatment of children with GHD. The target height does not seem to be as important as the birthweight in affecting AH in early-treated GHD children. In conclusion, the analysis of published data emphasizes the importance of an early diagnosis as the major factor influencing outcome. Consequently, the clinical features and an accurate physical examination are extremely important for this purpose. On the other hand, variables such as birthweight, height at onset of puberty, treatment regimens and aetiology of the defect also seem to be important factors in affecting the final height of treated GHD children.

References

1 Ranke MB: A note on adults with growth hormone deficiency. Acta Paediatr Scand Suppl 1987;331:80–82.
2 van der Werff ten Bosch JJ, Bot A: Growth of males with idiopathic hypopituitarism without growth treatment. Clin Endocrinol (Oxf) 1990; 32:707–717.
3 Rimoin DL, Merimee TJ, Rabinowitz D, McCusick VA: Genetic aspects of clinical endocrinology. Recent Prog Horm Res 1968;24: 365–437.
4 Raben MS: Treatment of a pituitary dwarf with human growth hormone. J Clin Endocrinol Metab 1958;18:901–903.
5 Burns EC, Tanner JM, Preece MA, Cameron N: Final height and pubertal development in 55 children with idiopathic growth hormone deficiency, treated for between 2 and 15 years with human growth hormone. Eur J Pediatr 1981;137:155–164.
6 Libber SM, Plotnick LP, Johanson AJ, Blizzard RM, Kwiterovich PO, Migeon CJ: Long-term follow-up of hypopituitary patients treated with human growth hormone. Medicine (Baltimore) 1990;69:46–55.
7 Frasier SD: Human growth hormone (hGH) in growth hormone deficiency. Endocr Rev 1983; 4:155–170.
8 Smith PJ, Hindmarsh PC, Brook CGD: Contibution of dose and frequency of administration to the therapeutic effect of growth hormone. Arch Dis Child 1988;63:491–494.
9 Kastrup KW, Christiansen JS, Andersen JK, Orskov H: Increased growth rate following transfer to daily s.c. administration from these weekly i.m. injections of hGH in growth hormone deficient children. Acta Endocrinol (Copenh) 1983;104:148–152.
10 Blethen SL, Baptista J, Kuntze J, Foley T, LaFranchi S, Johanson A: Adult height in growth hormone (GH)-deficient children treated with biosynthetic GH. J Clin Endocrinol Metab 1997;82:418–420.
11 Severi F: Final height in children with growth hormone deficiency. Horm Res 1995;43:138–140.
12 Braemswig JH, Schloesser H, Kiese K: Final height in children with growth hormone deficiency. Horm Res 1995;43:126–128.
13 Frisch H, Birnbacher R: Final height and pubertal development in children with growth hormone deficiency after long-term treatment. Horm Res 1995;43:132–134.
14 Rikken B, Massa GG, Wit JM: Final height in a large cohort of Dutch patients with growth hormone deficiency treated with growth hormone. Horm Res 1995;43:135–137.

15 Birnbacher R, Riedl S, Frisch H: Long-term treatment in children with hypopituitarism: Pubertal development and final height. Horm Res 1998;49:80–85.

16 Cutfield W, Lindberg A, Albertsson Wikland K, Chatelain P, Ranke MB, Wilton P: Final height in idiopathic growth hormone deficiency: The KIGS experience. Acta Paediatr 1999;(suppl 428):72–75.

17 Cacciari E, Cicognani A, Pirazzoli P, Zucchini S, Salardi S, Balsamo A, Cassio A, Pasini A, Carlà G, Tasinari D, Gualandi S: Final height of patients treated for isolated GH deficiency: Examination of 83 patients. Eur J Endocrinol 1997;137:53–60.

18 Arrigo T, DeLuca F, Maghnie M, Blandino A, Lombardo F, Messina MF, Wasniewska M, Ghizzoni L, Bozzola M: Relationship between neuroradiological and clinical features in apparently idiopathic hypopituitarism. Eur J Endocrinol 1998;139:84–88.

19 Cacciari E, Tassoni P, Cicognani A, Pirazzoli P, Salardi S, Balsamo A, Cassio A, Zucchini S, Colli C, Tassinari D, Tani G, Gualandi S: Value and limits of pharmacological and physiological tests to diagnose growth hormone (GH) deficiency and predict therapy response: First and second retesting during replacement therapy of patients defined as GH deficient. J Clin Endocrinol Metab 1997;79:1663–1669.

20 Cutfield WS, Lindberg A, Chatelain P, Price DA, Albertsson-Wikland K, Wilton P, Ranke MB: Final height following growth hormone treatment of idiopathic growth hormone deficiency in KIGS; in Ranke MB, Wilton P (eds): Growth Hormone Therapy in KIGS – 10 Years' Experience. Heidelberg, Barth, 1999, pp 93–109.

21 Hintz RL: Final height of growth hormone-treated patients with growth hormone deficiency: The North American experience. Acta Paediatr 1999;88(suppl 428):70–71.

22 Grumbach MM, Bin-Abbas BS, Kaplan SL: The growth hormone cascade: Progress and long-term results of gowth hormone tretment in growth hormone deficiency. Horm Res 1998;49(suppl 2):41–57.

23 Blethen SL, Compton P, Lippe BM, Rosenfeld RG, August GP, Johanson A: Factors predicting the response to growth hormone (GH) therapy in prepubertal children with GH deficiency. J Clin Endocrinol Metab 1993;76:574–579.

24 Job J-C: Early diagnosis and early treatment of growth hormone deficiency. Horm Res 1989;31:149–152.

25 Josefsberg Z, Bauman B, Pertzelan A, Laron Z: Greater efficiency of human growth hormone therapy in children below five years of age with growth hormone deficiency. A 5 year follow-up study. Horm Res 1987;27:126–133.

26 Arrigo T, De Luca F, Bernasconi S, Bozzola M, Cavallo L, Crisafulli G, Ghizzoni L, Maghnie M, Zecchino C: Catch-up growth and height prognosis in early treated children with congenital hypopituitarism. Horm Res 1995;44(suppl 3):26–31.

27 Arrigo T, Bozzola M, Cavallo L, Ghizzoni L, Maghnie M, Messina MF, Wasniewska M, De Luca F: Growth hormone deficient children treated from before two years old fail to catch-up completely within five years of therapy. J Pediatr Endocrinol Metab 1998;11:45–50.

28 Huet F, Carel JC, Nivelon JL, Chaussain JL: Long-term results of GH therapy in GH-deficient children treated before 1 year of age. Eur J Endocrinol 1999;140:29–34.

29 Rappaport R, Mugnier E, Limoni C, Crosnier H, Czernichow P, Leger J, Limal JM, Rochiccioli P, Soskin S: A 5-year prospective study of growth hormone (GH)-deficient children treated with GH before the age of 3 years. J Clin Endocrinol Metab 1997;82:452–456.

30 De Luca F, Maghnie M, Arrigo T, Lombardo F, Messina MF, Bernasconi S: Final height outcome of growth hormone-deficient patients treated since less than five years of age. Acta Paediatr 1999;85:1167–1171.

Hormone Research

Horm Res 2000;53(suppl 1):60–69

Metabolic Consequences of Growth Hormone Treatment in Paediatric Practice

Paul Saenger

Division of Pediatric Endocrinology, Albert Einstein College of Medicine/Montefiore Medical Center, Bronx, N.Y., USA

Key Words

Growth hormone · Carbohydrate tolerance · Water retention · Hyperlipidaemia · Carpal tunnel syndrome · Idiopathic intracranial hypertension

Abstract

No metabolic side-effects of clinical significance have been reported during a 5-year study of growth hormone (GH) therapy in children with GH deficiency, Turner syndrome, idiopathic short stature or chronic renal insufficiency. In particular, insulin levels increase but remain within the normal range, as do glucose and haemoglobin A_{1c}. A recent study showed that the effects of growth on insulin sensitivity in prepubertal children with idiopathic short stature represent the changes in carbohydrate tolerance observed during normal adolescence. Thus, GH treatment may lead to prolongation of the physiological state of insulin resistance observed in normal puberty. Insulin levels during the fasting state and 2 h after a standard glucose load showed no further rise after the first 3 years of continuous GH therapy. The hyperinsulinaemia observed during GH therapy may, therefore, amplify the anabolic effects of insulin on protein metabolism during puberty.

Introduction

There are currently almost 20,000 children on active treatment with growth hormone (GH) in the USA. With the widespread use of a product comes the inevitable reporting of various side-effects, some of which are serious and some relatively minor. In some patients, an undeniable cause-and-effect relationship has been established; in others, the role of GH is less convincing. After more than 30 years of use in children with GH deficiency (GHD), GH can be considered a well-tolerated drug with few side-effects. The purpose of this paper is to review the adverse effects and metabolic consequences of GH therapy.

The limited knowledge of the long-term effects of recombinant GH, together with the increasing number of patients treated, led health authorities in Europe, Japan, the USA and several other countries to organize long-term follow-up studies in a large cohort of patients. Two such ongoing studies document the use of GH and report adverse effects of therapy. The Pharmacia & Upjohn International Growth Study (KIGS) is based in Europe, Asia and Australia and now also in the USA, and the National Cooperative Growth Study (NCGS) has been established in the USA and Canada. Through these two

KARGER

Fax +41 61 306 12 34
E-Mail karger@karger.ch
www.karger.com

0301–0163/00/0537–0060$17.50/0

Accessible online at:
www.karger.com/journals/hre

Dr. P. Saenger
Division of Pediatric Endocrinology
Montefiore Medical Center, 111 East 210th Street
Bronx, NY 10467 (USA)
Tel. +1 718 920 4664, Fax +1 718 405 5609, E-Mail phsaenger@aol.com

agencies, thousands of children who are receiving GH for a wide range of indications are being monitored for the assessment of their growth and any signs of adverse effects of treatment.

GH Therapy and Carbohydrate Metabolism

Children have high turnover rates of the substrates that provide most of the energy required for the oxidative needs of body tissues: glucose, fatty acids and ketones. In the fasting state, childhood glucose homeostasis is a balance between hepatic production and utilization (4–6 mg/kg/min). Gluconeogenesis and glycogenolysis contribute about 75 and 25% to glucose production in the fasting state, respectively. The utilization of glucose in children is remarkably different to that in adults; the non-insulin-mediated uptake by the brain is high in children and contributes up to 80% of the total glucose demand compared with only 40% in adults. Insulin suppresses hepatic glucose production and stimulates peripheral glucose use, whereas glucagon, cortisol, GH and catecholamines act as counter-regulatory influences. Although hypoglycaemia usually does not occur in GH-deficient adults, fasting hypoglycaemia has been observed in children with GHD. The cause of low plasma glucose is a diminished hepatic output through decreased gluconeogenesis and/or abnormal glycogen mobilization [1]. This may be the result of a lack of antagonism of GH on insulin action. As a result of the relatively unopposed insulin action, there is also reduced secretion of insulin, but this compensatory mechanism does not appear to be sufficient to prevent occasional hypoglycaemia. Administration of GH to deficient children restores glucose production to normal levels [2].

It has long been accepted that GH exerts a 'diabetogenic' action in experimental animals [3]. Studies carried out with GH led to the observation that there is an early insulin-like action lasting about 2 h, involving enhanced glucose use and antilipolysis (this effect is eliminated by prior administration of GH), followed by a period of hyperglycaemia despite elevated insulin levels [4, 5]. Excess GH appears to cause insulin resistance both in peripheral tissues and in the liver with decreased peripheral glucose uptake and an increase in the hepatic glucose production [5]. Increased basal hepatic glucose production has even been noted in acromegalic individuals with normal glucose tolerance [6].

In hypopituitarism, GH therapy has been associated with normalization of glucose turnover and insulin secretion and has led to either unchanged or impaired responses to oral glucose administration [2, 7]. Ten prepubertal children with non-GH-deficient idiopathic short stature (ISS) showed an almost twofold rise in their insulin levels after 1 year of GH therapy, though none showed glucose intolerance. Interruption of treatment resulted in normalization of insulin secretion [8]. Hindmarsh et al. [9] measured glycaemic responses to GH therapy in short normal children. Fasting serum insulin concentrations increased from 5.4 ± 3.5 to 15.3 ± 8.7 mU/l at the end of the 1st year of treatment, results which are very similar to our own findings [10]. Fasting levels of serum insulin decreased slightly during the 2nd year of therapy. No postprandial insulin levels were reported by these investigators. They concluded that tolerance to the effects of GH on carbohydrate metabolism occurs.

In a group of 10 children with ISS, levels of plasma insulin increased during oral glucose tolerance testing, but an almost complete reversal of hyperinsulinism was observed within 1 year of discontinuation of GH therapy [11]. The dose of GH used in this group was 0.7 mg/kg/day, almost two to three times the standard dose normally used. In a large multicentre study, pubertal and prepubertal children with ISS displayed mean baseline and 2-hour postprandial glucose levels which remained unchanged throughout the 5-year treatment period. Although insulin levels showed a rise during the first 3 years of the study, they remained within the normal range, and after the 3rd year, no further rise was observed. Haemoglobin A_{1c} levels remained stable throughout the study [10]. It has been suggested that the effect of GH on insulin sensitivity in prepubertal children with ISS appears to mirror the changes in carbohydrate tolerance seen in normal adolescence [12]. The long-term implication of the prolongation of this physiological state of insulin resistance of normal puberty is unknown.

Mild insulin resistance is a feature of Turner syndrome, and this is frequently exaggerated by GH therapy. After 6–12 months of GH treatment, both first- and second-phase insulin responses increased during a hyperglycaemic clamp study. As the glucose infusion rates during the clamp did not change after GH administration, the increased insulin response indicates decreased insulin sensitivity [13]. In a study evaluating GH treatment, both alone and in combination with oxandrolone, in girls with Turner syndrome, there was a higher incidence of baseline pretreatment glucose intolerance which is a well-known phenomenon in Turner syndrome. There was no significant change in mean glucose or glucose tolerance in patients treated with GH alone during the 1st year of therapy, while insulin levels during glucose tolerance rose

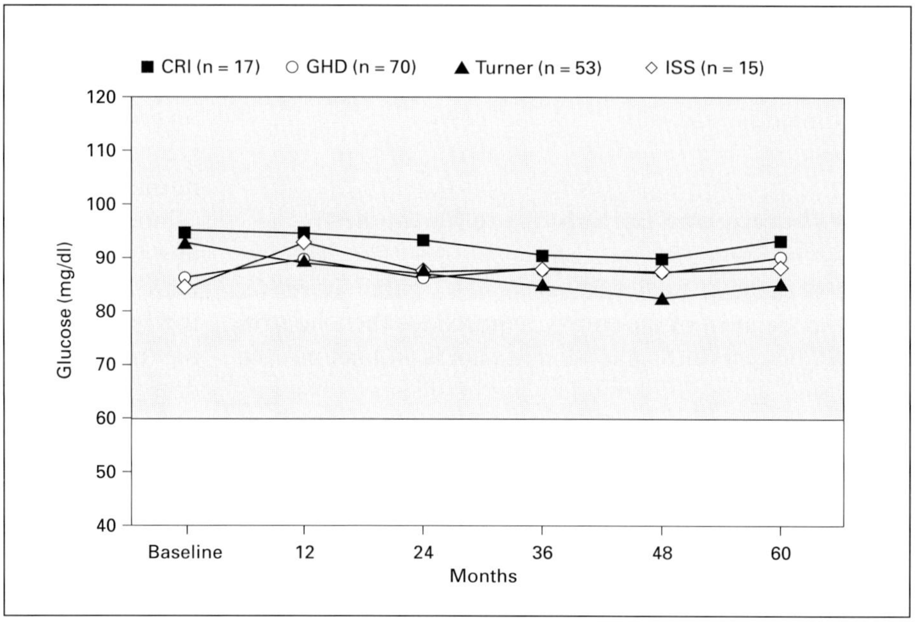

Fig. 1. Fasting glucose (mean) throughout the 5-year study period in children with GHD, Turner syndrome, ISS and CRI. The shaded band denotes the normal range. [From ref. 10, with permission].

[14]. In contrast, therapy with oxandrolone, both alone and in combination with GH, resulted in significant glucose intolerance and even higher insulin levels during the 1st year of therapy than GH therapy alone. In a recent Italian study, euglycaemic and hyperinsulinaemic clamp studies were performed in 6 prepubertal girls with Turner syndrome before and 6 and 12 months after starting GH therapy. They demonstrated that, although peripheral glucose use was not altered, there was a decrease in the hepatic insulin sensitivity [Monti and coworkers, personal commun.]. After discontinuation of GH treatment, a group of Turner syndrome patients who had elevated insulin levels while on therapy showed a progressive normalization of insulin levels [15].

Carbohydrate metabolism has been evaluated in children with chronic renal insufficiency (CRI) and compared with carbohydrate metabolism in various other growth disorders that are treated with GH. In one study, mean fasting and postprandial glucose remained unchanged throughout a 5-year period. Median fasting insulin levels rose from low normal levels into the normal range, with an average level of 10 mU/l at the end of 5 years. Mean haemoglobin A_{1c} levels of the 16 children with CRI were slightly elevated by the end of the study period compared with those of children with other growth disorders [10]. Similarly, other studies have demonstrated that, although no change in glucose tolerance was observed, an increase in insulin secretion occurred and persisted for the first 4 years in patients with CRI on conservative treatment and in transplant recipients [16, 17].

The stimulated insulin secretion remains a point of concern with regard to the possible development of glucose intolerance in the long term. This is particularly relevant in transplant recipients, for whom GH treatment appears as a third risk factor for insulin resistance, in addition to chronic renal failure itself and the diabetogenic effects of glucocorticoid treatment.

The rise in insulin levels during GH therapy in the various patient groups described in figures 1–5 is substantive, though levels remain within the normal range for postprandial insulin levels as defined by the reference laboratory used.

The clinical significance, if any, of these changes is unclear. Follow-up of patients without GHD treated for a similar duration with comparable doses of GH shows a decline in postprandial insulin levels after GH therapy is discontinued [16]. These data, therefore, do not support the concept that GH therapy of this duration will lead to beta cell exhaustion.

Heptulla et al. [18] recently reported that the effects of GH on insulin sensitivity in prepubertal children with ISS are nearly identical to the changes in carbohydrate tolerance observed during normal adolescence. GH treatment in ISS may thus lead to a prolongation of the physiological state of insulin resistance observed in normal puberty [19, 20]. The long-term implications of prolonged insulin

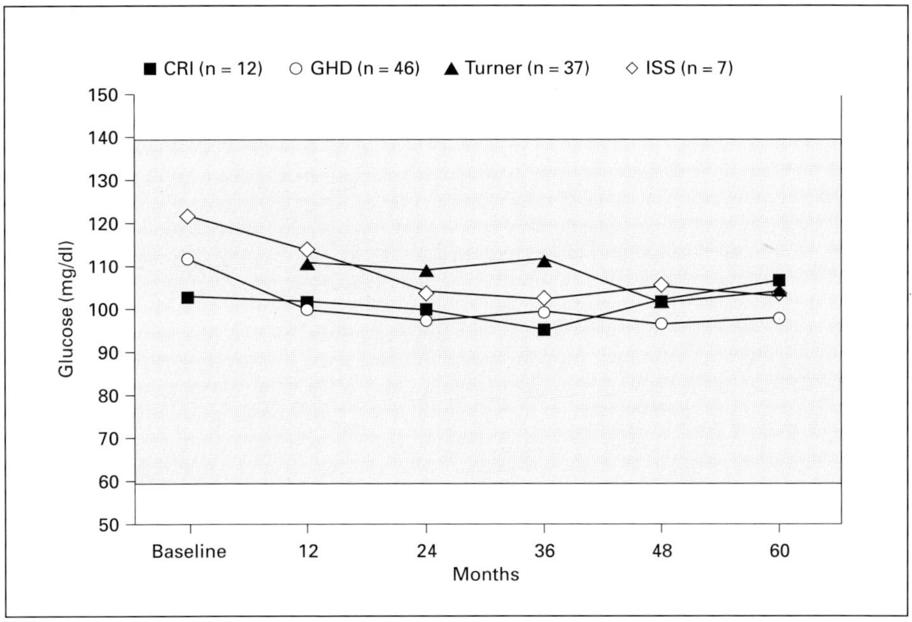

Fig. 2. Postprandial glucose (mean) levels throughout the 5-year study period in children with GHD, Turner syndrome, ISS and CRI. Baseline postprandial glucose levels are not available for the Turner syndrome patients. The shaded band denotes the normal range. [From ref. 10, with permission.]

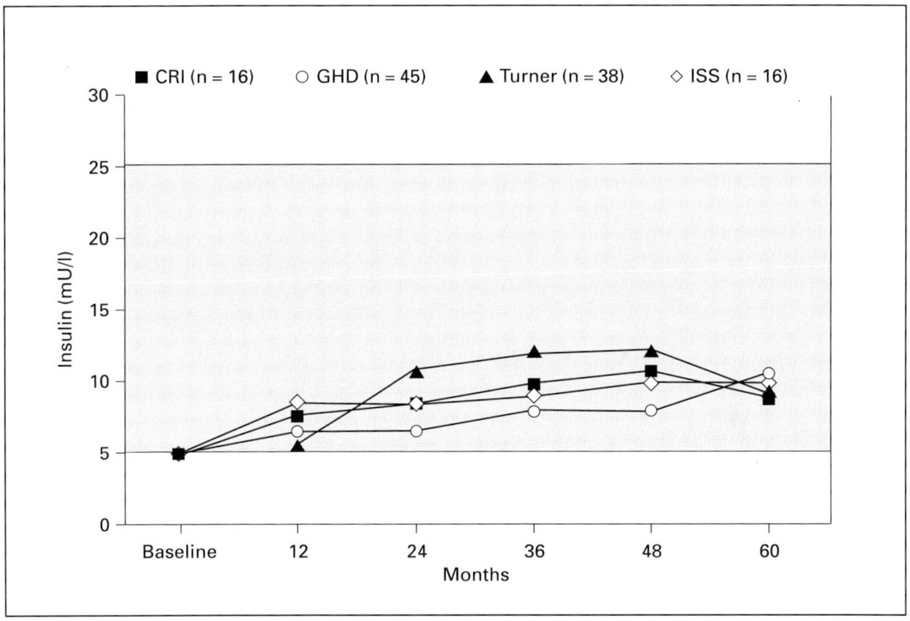

Fig. 3. Fasting insulin (median) levels throughout the 5-year study period in children with GHD, Turner syndrome, ISS and CRI. Baseline fasting insulin levels are not available for the Turner syndrome patients. The shaded band denotes the normal range. [From ref. 10, with permission.]

resistance are not known. It is reassuring, however, that despite uninterrupted GH treatment, the insulin levels, when measured in the fasting state and 2 h after a standard glucose load, showed no further rise after the first 3 years of GH therapy. More detailed analysis of the insulin resistance associated with puberty using euglycaemic and hyperglycaemic clamp techniques [19, 20] suggests that GH-induced insulin resistance spares hepatic action of insulin and is limited to peripheral glucose metabolism. The insulin resistance is thus associated with a greater fall in branched-chain and other essential amino acids. The hyperinsulinaemia observed during puberty or GH therapy may, therefore, amplify the anabolic effects of insulin on protein metabolism in puberty [20, 21].

It is currently not known whether the metabolic effects of GH are dependent on the mean serum value of GH over a given time period, the total GH circulating in plasma during that period (area under the curve), the number

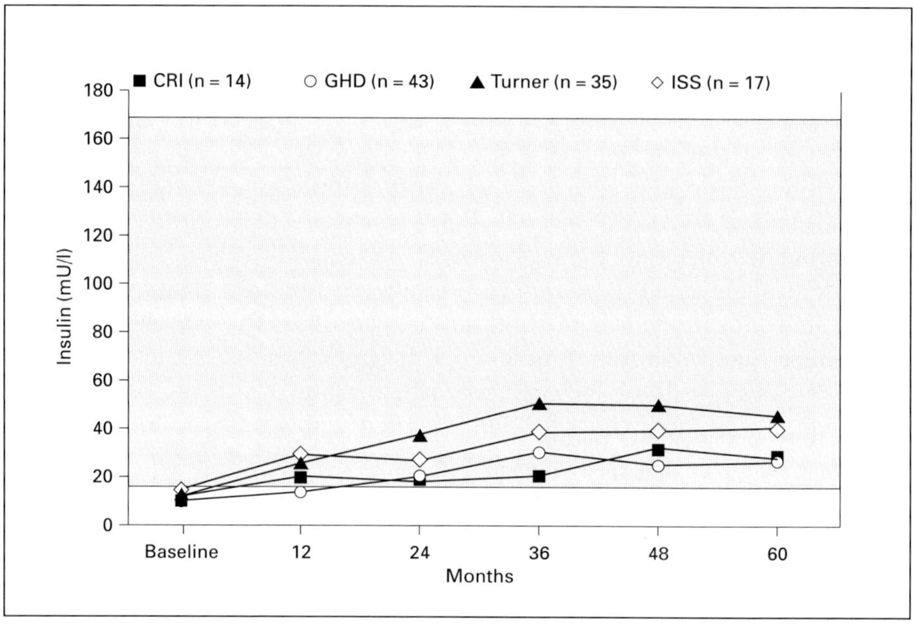

Fig. 4. Postprandial insulin (median) levels throughout the 5-year study period in children with GHD, Turner syndrome, ISS and CRI. The shaded band denotes the normal range. [From ref. 10, with permission.]

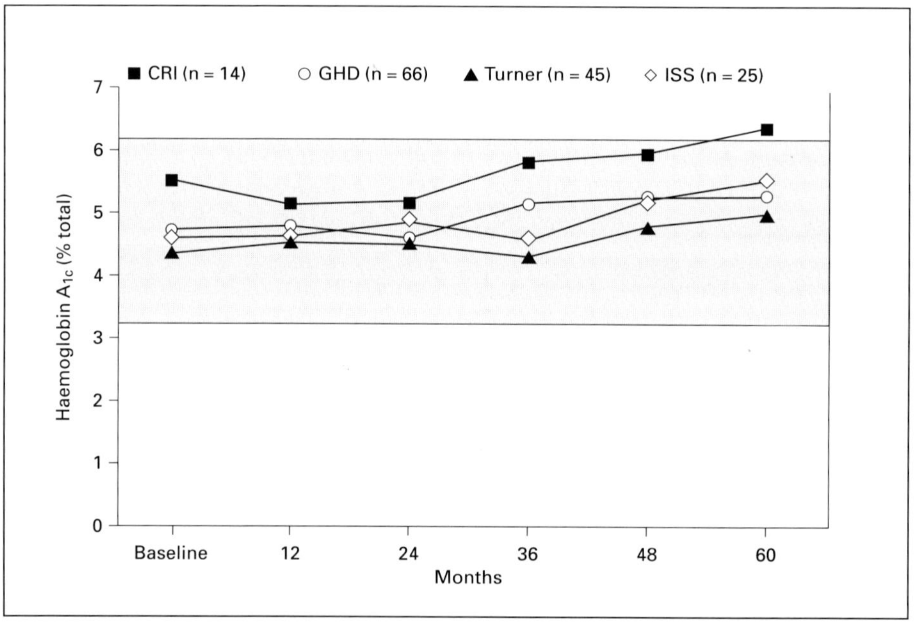

Fig. 5. Haemoglobin A_{1c} (mean) levels throughout the 5-year study period in children with GHD, Turner syndrome, ISS and CRI. Note assay changed 4/91. Reference range: before, 3.4–6.1%; after, 4.4–6.1%. The shaded band denotes the normal range. [From ref. 10, with permission.]

and/or amplitude of GH peaks achieved or the maximum GH levels achieved. Using the dosage schedule described in this study [10], peak GH levels as high as 32 ng/ml have been measured after subcutaneous injections [22]. This order of magnitude of GH levels is seen in 24-hour studies of GH dynamics in normal healthy control children [22]. In addition, insulin-like growth factor 1 concentrations or the concentration of insulin-like growth factor binding proteins may modulate the metabolic effects of GH [23].

The NCGS in the USA has monitored the safety of recombinant GH since 1985, and data have been collected from more than 20,000 children in the USA and Canada (table 1). Twenty-two children who developed diabetes mellitus during GH therapy were reported. In 9 children, the diabetes resolved when GH therapy was discontinued: 4 of these children were receiving steroids, and 2 had a prior history of carbohydrate intolerance. In the remaining 13 children, all but 3 had existing risk factors for developing diabetes. In summary, it was concluded that the incidence of permanent diabetes is similar to that reported in the normal North American adolescent population [24].

KIGS is the worldwide counterpart to the North American NCGS and includes more than 12,000 patients corresponding to more than 27,000 years of treatment [25]. The data suggest that there is a higher incidence of non-insulin-dependent diabetes mellitus (NIDDM) only in children who have predisposing risk factors and have subsequently received GH treatment. Although the incidence of NIDDM in normal children is not known, it is thought to be on the rise among various ethnic groups, particularly in the USA and Canada [26]. The incidence figure of 0.14% in the group treated for CNS tumours and the even larger figure of 0.55% in the group treated for leukaemia seem to be high. If there is an increased risk of diabetes after leukaemia treatment that does not include GH administration, this has not been reported. Thus, it is not possible to evaluate whether GH, with its known effect of decreasing sensitivity to insulin, plays a role in the aetiology of NIDDM, or whether other factors, such as the very high initial dose of corticosteroids, are involved. The 2 cases of insulin-dependent diabetes mellitus in 2,230 patient-years of GH treatment in patients with Turner syndrome, where glucose tolerance occurs during the natural course of the condition, do not represent a statistically increased risk. An even lesser risk is represented by 1 case of insulin-dependent diabetes mellitus in 11,972 patient-years of treatment comprising 4,053 children with idiopathic GHD, compared with as many as 28 per 100,000 reported in normal children in various countries [25].

Although safety results from the use of GH at the present dosage of 0.3 mg/kg/week in paediatric practice in the USA are reassuringly good, the use of GH in adults poses a different problem, with abnormal carbohydrate metabolism developing in elderly people, in obese patients, in postoperative states and in burn patients, even though doses are relatively smaller than those administered to children [27–30].

Although there appears to be no significant risk of children developing diabetes in the absence of other known risk factors, the long-term implications of insulin resistance and transient hyperinsulinaemia have yet to be elucidated. At this juncture, it is prudent to monitor the haemoglobin A_{1c} and glucose and insulin levels periodically while patients are on GH therapy. Individuals known to be at high risk of glucose intolerance, such as girls with Turner syndrome, post-chemotherapy patients and children with CRI, should be monitored very closely.

GH and Lipid Metabolism

Adipose tissue is a well-known target for GH action. GH is known to have a dual effect on lipids: the acute, short-lived effect is insulin-like, whereby lipolysis is acutely inhibited with a fall in the plasma free fatty acid (FFA) levels [31, 32]; the second and longer-lasting effect is the anti-insulin and lipolytic effect which results in a rise in plasma FFA and ketone levels [33, 34]. The lipolytic effect results in a decrease of total body adipose tissue during long-term GH treatment by increasing the hydrolysis of triglycerides, releasing FFAs and glycerol while decreasing FFA re-esterification [35]. This lipolytic action of GH allows lean body mass conservation at the expense of fat depots in obese individuals on a low-energy diet [36].

Lipo-atrophy may occur at the injection site, an adverse event only observed after repeated subcutaneous GH injections at the same site. In adults with GHD, high plasma cholesterol levels are accompanied by a rise in the concentrations of both low-density and very-low-density lipoproteins [37]. A raised plasma cholesterol (but not triglyceride) level has been observed in children with GHD, the cholesterol normalizing during long-term GH therapy [38].

After receiving 3 months of GH therapy, 7 GH-deficient children demonstrated a reduction in the lipid content of the abdominal subcutaneous adipocytes and a resultant redistribution of abdominal tissue stores to a more peripheral pattern. This site-specific change in adipose tissue distribution may be contributed by anatomically site-specific, GH-mediated changes in insulin responsiveness of adipose tissue. The authors [39] also demonstrated a decrease in de novo triglyceride synthesis with GH therapy. Fifteen children, some with GHD and some with normal GH levels, demonstrated an increase of their lean body mass with no change in their fat mass within 6 weeks of starting GH therapy [40]. In a study evaluating both carbohydrate and fat metabolism with GH treatment in non-GHD children, despite relatively large doses of GH (0.3 U/kg/day), FFA levels were unaffected, and the relative contributions of carbohydrate and lipid oxidation to the energy expenditure, measured by indirect calorimetry methods, were unchanged by GH therapy [11]. In a large study comparing the effects of GH on different groups of children, the modest changes that were observed in cholesterol and triglyceride levels were not statistically significant, and both levels remained in the normal range compared with age-matched standards [10].

GH treatment in a group of Turner syndrome patients receiving higher GH doses than usual had no effect on lipid metabolism [14, 41]. Although GH treatment did not affect the total plasma concentrations of triglyceride and cholesterol in children with CRI, recent evidence suggested that a transient increase in lipoprotein(a) may occur in some patients [42].

An atherogenic plasma lipoprotein profile is characteristic of CRI, and these alterations are perpetuated after renal transplantation by ciclosporin A and glucocorticoid medication. Hence, GH could theoretically contribute to early atherosclerosis in this group of patients, by both an increase in lipoprotein and the observed increase in insulin secretion [43].

During 30 months of administration of GH in Turner syndrome patients (age 6–15 years), the lipoprotein a levels did not change [43].

Although the typical fat deposition pattern in GH-deficient children is well known, and the GH effect on fat mass in such individuals appears to be beneficial, very little can be concluded regarding its effects on lipid metabolism, particularly in children with other underlying disorders.

Fluid, Electrolytes and Blood Pressure

The mean sodium and potassium concentrations showed no significant change during a 5-year study [10]. Transient sodium and water retention after pituitary GH administration was described 30 years ago [44]. In acromegaly, both intra- and extracellular fluid volumes are increased. Studies of GH-deficient adults using recombinant human GH yielded similar results [45]. In 6 of 24 adults, fluid retention was apparent after 4 weeks of GH treatment, as evidenced by swelling of hands and feet. In a study in adults [46] that sought to determine whether 3 weeks of treatment with GH might facilitate the preservation of nitrogen and accelerate the loss of body fat during dietary restriction, the injection of GH resulted in cessation of weight loss for at least 1 week in 5 of 8 patients; 5 patients developed mild oedema during the 1st week of treatment. Not surprisingly, after GH treatment was stopped, all patients experienced a water diuresis (mean 2.06 ± 1.25 kg) [46]. When higher doses of GH (10 mg/day for 7 days) were used in a study by Manson and Wilmore [47], water retention and oedema formation became even more prevalent.

Fluid retention has been demonstrated in children with GHD after both short- and long-term use of pituitary human GH by using antipyrine kinetics as an indirect measure of total body water [48]. Recent data by Ho and Weissberger [49] show that the administration of human GH at a dose of 0.1 mg/kg/day for 5 days in adult volunteers acutely leads to an activation of the renin-angiotensin axis which reached the maximum level 3–5 days after the first GH dose. The blood pressure did not change, however. Although fluid retention commonly accompanies GH excess, it is unlikely that the chronic volume expansion seen in acromegaly arises soley from GH-induced activation of mineralocorticoid secretion. It is well known that escape from sodium retention occurs promptly after protracted aldosterone elevation [49]. In a recent study [50] assessing the effects of 1 year of GH treatment on blood pressure, atrial natriuretic factor, aldosterone and plasma renin activity, no changes compatible with sodium retention were seen. The relationship between transient sodium retention and pseudotumour cerebri accompanying GH administration remains to be explored [24].

Idiopathic Intracranial Hypertension (Pseudotumour cerebri)

The occurrence of pseudotumour cerebri or idiopathic intracranial hypertension (IIH), a benign condition with raised intracranial pressure and papilloedema, has been attributed to fluid shifts; it is usually encountered, however, in patients who are also receiving glucocorticoids or who have other predisposing conditions. Although IIH is believed to be rare in children, it has been described in children with CRI and in association with replacement therapy in primary hypothyroidism. In children with GHD, it is possible that rapid correction of long-standing hormone deficiency results in intracerebral fluid shifts and the development of IIH. Of the 20 reported cases of IIH, 7 patients had GHD, and 8 had CRI [51, 52]. The remaining patients had Turner syndrome and Prader-Willi syndrome, but some patients had no underlying disorder. Obesity, CRI and other metabolic and endocrine disorders, such as hypothyroidism, may themselves increase the risk of mild IIH, though the use of higher doses of GH may also play a part.

Most studies that have reported water and salt retention describe IIH as a temporary self-limiting phenomenon occurring during the first few weeks of therapy. Nevertheless, the patient and parents should be instructed to report any untoward weight gain, feelings of tightness and especially headaches, photophobia and visual changes which may occur with raised intracranial pressure. Fundoscopy is recommended before treatment and at follow-up visits. Fortunately, almost all reports of IIH have also shown that stopping treatment reverses the process, and subsequently restarting of GH therapy does not appear to cause a recurrence.

Table 1. Recommendations for management of possible GH side-effects [modified from ref. 24]

Symptom	Possible cause	Evaluation	Management
Metabolic effects			
Polyuria, polydipsia	diabetes mellitus	urine, glucose, serum glucose and insulin	address risk factors, diet and insulin
Weight gain, swelling	oedema	using desmopressin (?)	alter desmopressin dose; stop GH treatment; if symptoms abate, resume GH at lower dose (?)
Headache	pseudotumour cerebri	fundoscopy and neurological evaluation	if papilloedema, stop GH; rule out space-occupying lesion by MRI; treat increased intracranial pressure
Paraesthesias	carpal tunnel syndrome	nerve conduction studies	stop GH; if symptoms abate, resume GH at lower dose
Growth-related effects			
Limb, hip or knee pain	slipped capital femoral epiphyses	radiographs	stop GH until orthopaedic evaluation; if high-titre antibodies, change preparation after washout period
Poor growth	antibodies (?)	review dose, methods, compliance, bone age, nutrition and thyroid status and then consider antibodies	
Other effects			
Previous neoplasms	recurrence or new neoplasm	evaluate for recurrence before and during GH therapy	stop GH
Naevi	rapid growth	examine naevi as routine	biopsy suspicious lesions

Carpal Tunnel Syndrome

Carpal tunnel syndrome, considered to be secondary to soft-tissue growth or oedema, is less common in children compared with adults: only 5 children in the NCGS and 2 children in non-NCGS studies are reported to have this syndrome [24]. This complication appears to be dose related.

Conclusions

Careful monitoring has not revealed any discernible metabolic side-effects of clinical significance during this long-term study of GH therapy in children with GHD, ISS, CRI and Turner syndrome [10]. In particular, insulin levels increase but remain within the normal range, as do glucose and haemoglobin A_{1c}. Table 1 presents management recommendations for possible GH side-effects, spanning both metabolic and growth-related effects. The metabolic side-effects of long-acting GH [53] or markedly increased doses of GH, which have been explored for pubertal children with GHD [54], have not yet been fully evaluated. A recent survey of 23,333 children who received GH treatment revealed abnormal glucose metabolism in 85 (0.36%). The incidence of type-1 diabetes mellitus was not increased. The incidence of type-2 diabetes mellitus was increased. The authors suggest that GH treatment may accelerate the disorder in predisposed patients [55]. Although GH treatment in children appears to have a good safety profile, continued surveillance is necessary [24, 55–57].

References

1 Haymond MW, Karl I, Weldon VV, Pagliara AS: The role of growth hormone and cortisone on glucose and gluconeogenic substrate regulation in fasted hypopituitary children. J Clin Endocrinol Metab 1976;42:846–856.

2 Bougnères PF, Artavia-Loria E, Ferre P, Chaussain JL, Job JC: Effects of hypopituitarism and growth hormone replacement therapy on the production and utilization of glucose in childhood. J Clin Endocrinol Metab 1985;61:1152–1157.

3 Houssay BA: The hypophysis and metabolism. N Engl J Med 1936;241:961.

4 Rizza R, Miles J, Verdonk C: Insulin sensitivity in man: A method for single day dose-response assessment using the euglycemic glucose insulin clamp technique. Clin Res 1980;28:409–414.

5 Davidson MB: Effect of growth hormone on carbohydrate and lipid metabolism. Endocr Rev 1987;8:115–131.

6 Karlander S, Vranic M, Efendic S: Increased glucose turnover and glucose cycling in acromegalic patients with normal glucose tolerance. Diabetologia 1986;29:778–783.

7 Lippe BM, Kaplan SA, Golden MP, Hendricks SA, Scott ML: Carbohydrate tolerance and insulin receptor binding in children with hypopituitarism: Responses after acute and chronic human growth hormone administration. J Clin Endocrinol Metab 1981;53:507–513.

8 Walker J, Chaussain JL, Bougnères PF: Growth hormone treatment of children with short stature increases insulin secretion but does not impair glucose disposal. J Clin Endocrinol Metab 1989;69:253–258.

9 Hindmarsh PC, Pringle PJ, Di Silvio L, Brook CGD: Effects of 3 years of growth hormone therapy in short normal children. Acta Paediatr Scand Suppl 1990;366:6–12.

10 Saenger P, Attie KM, Di Martino-Nardi J, Fine RN: Carbohydrate metabolism in children receiving growth hormone for 5 years: Chronic renal insufficiency compared with growth hormone deficiency, Turner syndrome, and idiopathic short stature. Genentech Collaborative Group. Pediatr Nephrol 1996;10:261–263.

11 Lesage C, Walker J, Landier F, Chatelain P, Chaussain JL, Bougnères PF: Near normalization of adolescent height with growth hormone therapy in very short children without growth hormone deficiency. J Pediatr 1991;119:29–34.

12 Boulware SD, Caprio S, Jones TW, Sherwin RS, Tamborlane WV: Metabolic alterations induced by growth hormone treatment mirror changes during normal adolescence (abstract). Pediatr Res 1990;27:72.

13 Caprio S, Boulware SD, Press M, Sherwin RS, Rubin K, Carpenter TO, Plewe G, Tamborlane WV: Effect of growth hormone treatment on hyperinsulinemia associated with Turner syndrome. J Pediatr 1992;120:238–243.

14 Wilson DM, Frane JW, Sherman B, Genentech Turner's Collaborative Group: Carbohydrate and lipid metabolism in Turner syndrome: Effect of therapy with growth hormone, oxandrolone, and a combination of both. J Pediatr 1988;112:210–217.

15 Saenger P, Wesoly S, Wasserman E, Glickstein J, Issenberg H: Safety aspects of GH therapy in Turner syndrome (TS): No evidence for ventricular hypertrophy (VH) and normalization of insulin levels after discontinuation of GH (abstract). Pediatr Res 1996;39:98.

16 Tönshoff B, Haffner D, Mehls O, Dietz M, Ruder H, Blum WF, Heinrich U, Stover B. Efficacy and safety of growth hormone treatment in short children with renal allografts: Three year experience. Members of the German Study Group for Growth Hormone Treatment in Children with Renal Allografts. Kidney Int 1993;44:199–207.

17 Haffner D, Wuhl E, Tönshoff B, Mehls O, the German Study Group for Growth Hormone Treatment in Chronic Renal Failure: Growth hormone (GH) treatment in short children with chronic renal failure: 5-year experience (abstract). Nephrol Dial Transplant 1994;9:960–961.

18 Heptulla RA, Boulware SD, Caprio S, Silver D, Sherwin RS, Tamborlane WV: Decreased insulin sensitivity and compensatory hyperinsulinemia after hormone treatment in children with short stature. J Clin Endocrinol Metab 1997;82:3234–3238.

19 Caprio S, Plewe G, Diamond MP, Simonson DC, Boulware SD, Sherwin RS, Tamborlane WV: Increased insulin sensitivity in puberty: A compensatory response to reductions in insulin sensitivity. J Pediatr 1989;114:963–967.

20 Amiel SA, Caprio S, Sherwin RS, Plewe G, Haymond MW, Tamborlane WV: Insulin resistance of puberty: A defect restricted to peripheral glucose metabolism. J Clin Endocrinol Metab 1991;72:277–282.

21 Kearns GL, Kemp SF, Findik JP: Single and multiple dose pharmacokinetics of methionyl growth hormone in children with idiopathic growth hormone deficiency. J Clin Endocrinol Metab 1991;72:1148–1156.

22 Rose SR, Rose JL, Uriarte M, Barnes KM, Cassorla FG, Cutler GB Jr: The advantage of measuring stimulated as compared with spontaneous growth hormone levels in diagnosis of growth hormone deficiency. N Engl J Med 1988;319:201–207.

23 Martha PM Jr, Reiter EO, Davila N, Shaw MA, Holcombe JH, Baumann G: The role of body mass in response to growth hormone therapy. J Clin Endocrinol Metab 1992;75:1470–1473.

24 Blethen SL, Allen DB, Graves D, August G, Moshang T, Rosenfeld R: Safety of recombinant deoxyribonucleic acid-derived growth hormone: The National Cooperative Growth Study experience. J Clin Endocrinol Metab 1996;81:1704–1710.

25 Wilton P, Chatelain P, Cutfield W, Ranke MB, Price DA, Albertsson-Wikland K on behalf of the KIGS International Board: Long-term safety of growth hormone (GH0) treatment in Children in KIGS (abstract). The Endocrine Society Annual Meeting, San Diego, 1999, p 111.

26 Grinstein G, Aponte L, Vuguin P, Saenger P, Di Martino-Nardi J: Presentation and 5-year follow-up of non-insulin dependent diabetes mellitus (NIDDM) in minority youth (abstract). The Endocrine Society Annual Meeting, San Diego, 1999, p 97.

27 Belcher HJ, Mercer D, Judkins KC, Shalaby S, Wise S, Marks V, Tanner NS: Biosynthetic human growth hormone in burned patients: A pilot study. Burns 1989;15:99–107; Burns 1989;15:273 (erratum).

28 Marcus R, Butterfield G, Holloway L, Gilliland L, Baylink DJ, Hintz RL, Sherman BM: Effects of short-term administration of recombinant human growth hormone to elderly people. J Clin Endocrinol Metab 1990;70:519–527.

29 Snyder DK, Underwood LE, Clemmons DR: Anabolic effects of growth hormone in obese diet restricted subjects are dose dependent. Am J Clin Nutr 1990;52:431–437.

30 Ziegler TR, Young LS, Ferrari-Baliviera E, Demling RH, Wilmore DW: Use of human growth hormone combined with nutritional support in a critical care unit. JPEN 1990;14:574–581.

31 Goodman HM: Antilipolytic effects of growth hormone. Metabolism 1970;19:849–855.

32 Birnbaum RS, Goodman HM: Studies on the mechanism of the antilipolytic effects of growth hormone. Endocrinology 1976;99:1336–1345.

33 Swislocki NI, Szego CM: Acute reproduction of plasma nonesterified fatty acids by growth hormone in hypophysectomized and Houssay rats. Endocrinology 1965;76:665–672.

34 Chernick SS, Clark CM Jr, Gardiner RJ, Scow RO: Role of lipolytic and glucocorticoid hormones in the development of diabetic ketosis. Diabetes 1972;21:946–954.

35 Van Vliet G, Bosson D, Craen M, Du Caju MV, Malvaux P, Vanderschueren-Lodeweyckx M: Comparative study of the lipolytic potencies of pituitary-derived and biosynthetic human growth hormone in hypopituitary children. J Clin Endocrinol Metab 1987;65:876–879.

36 Clemmons DR, Snyder DK, Williams R, Underwood LE: Growth hormone administration conserves lean body mass during dietary restriction in obese subjects. J Clin Endocrinol Metab 1987;64:878–883.

37 Merimee IJ, Hollander W, Fineberg SE: Studies of hyperlipidemia in hGH-deficient state. Metabolism 1972;21:1053–1061.

38 Winter RJ, Thompson RG, Green OC: Serum cholesterol and triglycerides in children with growth hormone deficiency. Metabolism 1979;28:1244–1249.

39 Rosenbaum M, Gertner J, Leibel RL: Effects of systemic growth hormone administration on regional adipose tissue distribution and metabolism in GH deficient children. J Clin Endocrinol Metab 1989;69:1274–1281.
40 Gregory JW, Greene SA, Jung RT, Scrimgeour CM, Rennie MJ: Changes in body composition and energy expenditure after six weeks' growth hormone treatment. Arch Dis Child 1991;66:598–602.
41 Stahnke N, Stubbe P, Keller E, Zeisel HJ, Serono Group Hamburg: Effects and side-effects of GH plus oxandrolone in Turner syndrome; in Ranke MB, Rosenfeld RG (eds): Turner Syndrome: Growth Promoting Therapies. Amsterdam, Excerpta Medica, 1991, pp 241–247.
42 Querfeld U, Dopper S, Gradehand A, Kiencke P, Wahn F, Zeisel HJ: Long-term treatment with growth hormone has no persisting effect on lipoprotein(a) in patients with Turner's syndrome. J Clin Endocrinol Metab 1994;84:967–970.
43 Querfeld U, Haffner D, Wuhl E, Wingen AM, Walter K, Friedrich B, Michalk DV, Mehls O: Treatment with growth hormone increases lipoprotein(a) serum levels in children with chronic renal insufficiency. Eur J Pediatr 1996;155:913.
44 Hennemann PH, Forbes AP, Moldawer M, Dempsey EF, Carroll EL: Effects of human growth hormone in man. J Clin Invest 1960;39:1223–1238.
45 Salomon F, Cuneo RC, Hesp P, Sonksen PH: The effects of treatment with recombinant growth hormone on body composition and metabolism in adults with growth hormone deficiency. N Engl J Med 1989;321:1797–1803.
46 Clemmons DR, Snyder DK, Williams R, Underwood LE: Growth hormone administration conserves lean body mass during dietary restriction in obese subjects. J Clin Endocrinol Metab 1987;64:878–883.
47 Manson JM, Wilmore DW: Positive nitrogen balance with human growth hormone and hypocaloric intravenous feeding. Surgery 1986;100:188–196.
48 Rifkind AB, Saenger P, Levine LS, Pareira J, New MI: Effects of growth hormone on antipyrine kinetics in children. Clin Pharmacol Ther 1981;30:127–132.
49 Ho KY, Weissberger AJ: The antinatriuretic action of biosynthetic human growth hormone in man involves activation of the renin-angiotensin system. Metabolism 1990;39:133–137.
50 Di Martino-Nardi J, Wesoly S, Schwartz L, Saenger P: Lack of clinical evidence of sodium retention in children with idiopathic short stature treated with recombinant growth hormone. Metabolism 1993;42:730–734.
51 Malozowski S, Tanner LA, Wysowski D, Fleming GA: Growth hormone, insulin-like growth factor 1, and benign intracranial hypertension. N Engl J Med 1993;329:665–666.
52 Malozowski S, Tanner LA, Wysowski DK, Fleming GA, Stadel BV: Benign intracranial hypertension in children with growth hormone deficiency treated with growth hormone. J Pediatr 1995;126:996–999.
53 Saenger P, Mauras N, Reiter EO, Baptista J, Attie KM, the Genentech Collaborative Study Group: Effects of high dose rhGH therapy in adolescent children with GH deficiency: A randomized multicenter study (abstract). Horm Res 1998;50(suppl 3):10.
54 Silverman BL, Kemp SF, Attie KM, Ford KM: Treatment of childhood GH deficiency with a sustained-release growth hormone given once or twice monthly (abstract). Horm Res 1999;51 (suppl 2):16.
55 Cutfield WS, Wilton P, Bennmarker H, Albertsson-Wikland K, Chatelain P, Ranke MB, Price DA: Incidence of diabetes mellitus and impaired glucose tolerance in children and adolescents receiving growth-hormone treatment. Lancet 2000;355:610–613.
56 Saenger P, Attie KM, DiMartino-Nardi J, Hintz R, Frahm L, Frane J, the Genentech Collaborative Study Group: Metabolic consequences of 5-year growth hormone (GH) therapy in children treated with GH for idiopathic short stature. J Clin Endocrinol Metab 1998;83:3115–3120.
57 Banerjee S, Saenger P: Metabolic side effects and adverse effects of growth hormone therapy; in Kelnar CJH, Savage MO, Stirling HF, Saenger P (eds): Growth Disorders, Pathophysiology and Treatment. London, Chapman and Hall, 1998, pp 745–764.

Hormone Research

Horm Res 2000;53(suppl 1):70–76

Growth Hormone and Neurofibromatosis

Simon J. Howell[a] Patrick Wilton[b] Anders Lindberg[b] Stephen M. Shalet[a]

[a]Department of Endocrinology, Christie Hospital NHS Trust, Manchester, UK;
[b]KIGS/KIMS, Pharmacia & Upjohn, Stockholm, Sweden

Key Words
Neurofibromatosis · Growth hormone therapy, neurofibromatosis · Growth hormone deficiency · Intracranial tumour

Abstract
Data collected from 102 neurofibromatosis (NF) children with growth hormone (GH) deficiency (GHD) who were receiving GH replacement therapy were reviewed to assess the efficacy and safety of GH therapy in this condition. GH was administered at a mean dose of 0.18 mg/kg/week. During the 1st year the median height velocity increased significantly from 4.2 cm/year before treatment to 7.1 cm/year, and the median height standard deviation score increased from –2.4 to –1.9. The response to therapy, however, was not as good as that observed in patients with idiopathic GHD. GH therapy did not influence the progression of any of the features of NF, including intracranial tumours, and was not associated with an excess of other adverse events. We conclude that GH treatment of NF patients with GHD is beneficial in terms of improved growth rate and is well tolerated.

Introduction

Neurofibromatosis (NF) is one of the most common autosomal dominant disorders, with a frequency of 1 in 3,000 [1]. The gene for the most common form of the disorder, NF type I, has been located in the pericentric region of chromosome 17 [2]. The disease may manifest with many different features and has a markedly variable expression. The most common signs are café-au-lait spots, neurofibromas in the skin and deeper peripheral nerves and freckling in hyperpigmented areas such as the axillae and inguinal regions. Malignancy accounts for a major portion of the severe morbidity and mortality in this disease. Intracranial tumours, most commonly gliomas of the optic pathways, occur in 15–20% of NF patients [3, 4]. Other neoplasms, such as neurofibrosarcomas, rhabdomyosarcomas and several types of leukaemia, also occur more commonly in NF.

Endocrine abnormalities are also more prevalent in individuals with NF. Precocious puberty occurs in 3–5% of NF children [4, 5] and is almost always associated with an intracranial tumour (usually an optic tract glioma) or its treatment. Short stature has long been known to be a feature of NF [1], and it is clear that growth hormone (GH) deficiency (GHD) is more common in these patients, though the exact incidence has not been reported. The aetiology of GHD is not clear, but it is much more common in the presence of an intracranial tumour, and in

KARGER

Fax + 41 61 306 12 34
E-Mail karger@karger.ch
www.karger.com

0301–0163/00/0537–0070$17.50/0

Accessible online at:
www.karger.com/journals/hre

Prof. S.M. Shalet
Department of Endocrinology, Christie Hospital NHS Trust
Wilmslow Road
Manchester M20 4BX (UK)
Tel. +44 161 446 3667, Fax +44 161 446 3772, E-Mail Mmassey@picr.man.ac.uk

some cases is clearly related to treatment of these tumours with radiotherapy [6]. As children with NF have a greatly increased risk of malignancy, there has been some concern about the safety of GH treatment in this situation. There are no published data on the efficacy and safety of GH therapy in children with NF. We have, therefore, reviewed data in a cohort of 102 children with both NF and biochemical evidence of GHD who had received GH replacement therapy.

Patients and Methods

The data were collected retrospectively from the KIGS (Kabi International Growth Study) database (Pharmacia & Upjohn International Growth database) which was set up in 1987 to monitor the progress of children treated with recombinant GH (Somatonorm® or Genotropin®; Pharmacia & Upjohn, Stockholm, Sweden). A total of 102 patients with NF were identified. The diagnosis in all instances was made by the investigator in charge of the patient's care. A total of 66 of the patients were classified on the database primarily as NF, while 36 had a history of NF but were classified as having either idiopathic GHD (n = 6) or GHD in association with a tumour (28 with a glioma/astrocytoma, 1 with a sarcoma and 1 with an embryonal rhabdomyosarcoma). Sixteen of the 66 NF patients also had a tumour (14 a glioma/astrocytoma, 1 an alternative cranial tumour and 1 a rhabdomyosarcoma). Therefore, a total of 43 patients had an intracranial tumour, of which 30 were optic gliomas. Thirty-five patients received cranial radiotherapy (34 with glioma/astrocytoma and 1 with sarcoma), 2 of whom also received spinal irradiation. Twenty-one of the glioma/astrocytoma patients were also treated with chemotherapy, and 6 underwent surgery.

Data were available for 102 patients prior to GH therapy, 83 after 1 year of treatment, 57 after 2 years and 41 after 3 years. The smaller number of patients in whom data were available after 2 and 3 years reflects the fact that a proportion of the children had received GH replacement for less than 1, 2 or 3 years, respectively, at the time of data analysis. However, there were no drop-outs from the analysis. The median duration of GH treatment was 2.7 years, and 10% of patients were treated for more than 6 years. The height standard deviation score (SDS) was calculated using standards from Tanner and Whitehouse [7], and birth weight and length were calculated using standards from Niklasson et al. [8]. The target height (cm) was estimated in boys as 0.5 × (mother's height + father's height +13) and in girls as 0.5 × (mother's height + father's height – 13).

The response to GH therapy was assessed by both the height velocity (cm/year) calculated over 1 year and the change in height SDS (Δ height SDS). Entry into puberty was defined as breast stage 2 for girls and testicular volumes of 4 ml for boys and was recorded in 78 children at the start of GH therapy.

Results

The diagnosis of GHD was based on auxological criteria (height SDS and height velocity), coupled in most cases with the GH response to standard stimulation tests (most commonly the insulin tolerance test or the arginine stimulation test). Two tests were performed in 60 patients, one test was used in 31 patients, and no test was recorded in the remaining 11 patients. The most likely explanation for the lack of a test result in these 11 patients is that a GH stimulation test was performed, but the result was not reported to KIGS. The median GH peak was reduced in keeping with the diagnosis of GHD, but was significantly lower in patients with a tumour than in those with no tumour (3.0 vs. 4.6 μg/l; $p < 0.001$). Ninety-two per cent and 80% of the GH-tested patients with a cranial tumour had a peak GH response to all tests below 10 and 5 μg/l, respectively, while figures of 81 and 56%, respectively, were obtained in the non-tumour group.

Additional pituitary hormone deficiencies occurred in a minority of patients. Two patients in the non-tumour group were thyroid-stimulating hormone (TSH) deficient, with 1 also being adrenocorticotrophic hormone deficient. Of the patients with tumours, 26% were TSH deficient, 5% adrenocorticotrophic hormone deficient, 22% gonadotrophin deficient (only girls above 14 and boys above 15 years of age considered) and 4% antidiuretic hormone deficient. Twelve patients with precocious puberty were reported, all of whom had a cranial tumour. At the start of therapy, 68% of the children were prepubertal. After 1 year of GH therapy, 55% of the patients had entered puberty, and 62 and 83% were pubertal by the end of the 2nd and 3rd years of treatment, respectively. At all times there was a higher proportion of patients in the tumour group who had entered puberty than in the non-tumour group (39 vs. 27% before treatment, 70 vs. 44% after 1 year, 67 vs. 59% after 2 years and 91 vs. 80% after 3 years). As there was no difference in age between the two groups, this is probably a genuine reflection of the known association of precocious puberty in those children with a history of a tumour.

Patient characteristics at the start of GH therapy are shown in table 1. The median height SDS and median height velocity are both reduced at –2.4 and 4.2 cm/year, respectively. The children in the tumour group were taller with a pre-treatment median height SDS of –1.2 compared with –2.8 in the non-tumour group ($p < 0.0001$). This difference persisted when the height was adjusted for parental height (height SDS – target height SDS –0.8 vs. –1.6; $p < 0.001$) and even when only prepubertal children were considered (height SDS –1.2 vs. –2.7). The bone age was retarded with a median delay of 1.6 years but this was less marked in the tumour group (median bone age delay 1.1 years versus 1.9 years in the non-tumour group).

Table 1. Characteristics of 102 NF patients prior to GH therapy

	Birth-weight SDS	Birth-length SDS	Target height SDS	Age at diagnosis of NF, years	At start of GH treatment				
					age, years	height SDS	height-target height SDS	height velocity cm/year	bone age delay years
n	89	59	95	36	102	102	95	59	61
Median	−0.3	−0.3	−0.8	5.6	10.4	−2.4	−1.3	4.2	1.6
10th to 90th centile	−2.0 to +1.0	−1.7 to +1.3	−2.2 to +0.6	0.9–10.7	6.2–13.9	−3.8 to +0.0	−3.0 to +0.1	1.7–6.4	0.1–3.4

Table 2. Results of treatment of whole cohort expressed as medians (10th to 90th centile)

	Before treatment	Year 1	Year 2	Year 3
n	102	83	57	41
Age, years	10.4 (6.2–13.9)	11.6 (7.2–14.9)	13.0 (8.4–15.9)	14.2 (9.3–16.8)
Bone age delay, years	1.6 (0.1–3.4)	2.2 (−0.3 to +4.4)	1.7 (0.0–3.8)	1.8 (−0.6 to +3.2)
Height velocity, cm/year	4.2 (1.7–6.4)	7.1 (4.6–10.0)	5.7 (2.9–8.3)	5.7 (2.6–7.9)
Height SDS	−2.4 (−3.8 to 0.0)	−1.9 (−3.5 to 0.0)	−1.7 (−3.4 to +0.1)	−1.8 (−4.0 to +0.4)
Height SDS-target height SDS	−1.3 (−3.0 to +0.1)	−1.0 (−3.2 to + 0.1)	−0.9 (−2.9 to +0.4)	−0.7 (−3.2 to +0.4)
Change in height SDS	not applicable	0.3 (−0.2 to +0.7)	0.1 (−0.2 to +0.6)	−0.1 (−0.3 to +0.4)

Table 3. Results of treatment of 20 prepubertal children expressed as medians (10th to 90th centile)

	Pretreatment	Year 1	Year 2
Age, years	8.7 (5.8–11.3)	9.7 (6.7–12.4)	10.7 (7.7–13.4)
Bone age delay, years	2.3 (1.8–3.4)[a]	2.7 (0.1–3.8)[b]	2.1 (1.5–3.8)[a]
Height velocity, cm/year	4.1 (3.0–5.9)[b]	7.7 (5.5–9.8)	5.7 (4.1–8.4)
Height SDS	−2.4 (−3.3 to −0.9)	−1.8 (−2.6 to −0.2)	−1.7 (−2.6 to 0.0)
Height SDS-target height SDS	−1.2 (−3.1 to −0.3)[c]	−0.6 (−2.6to −0.2)	−0.3 (−2.6 to 0.0)
Change in height SDS	not applicable	0.4 (0.0–0.8)	0.1 (−0.2 to +0.5)

[a] n = 12; [b] n = 11; [c] n = 19.

GH replacement was administered at a mean dose of 0.18 mg/kg/week with the tumour group receiving a slightly lower dose than the non-tumour group (0.17 vs. 0.19 mg/kg/week). Table 2 illustrates the response to GH therapy. The median height velocity increased significantly from 4.2 cm/year before treatment to 7.1 cm/year during the 1st year and remained above the baseline value during the 2nd and 3rd years of therapy. The median height SDS increased from −2.4 to −1.9 in the first year of GH and remained relatively stable thereafter. There was no significant difference in growth response to treatment between the tumour group and the non-tumour group or between those who had received cranial radiotherapy and those who had not.

The growth response to GH therapy in patients with idiopathic GHD has been found to depend on a number of factors [9]. First-year height velocity is positively correlated with target height SDS, dose of GH and injection frequency and negatively correlated with age and pretreatment height SDS. A model has been developed in

KIGS that can predict the 1st-year height velocity of a patient based on these characteristics and, therefore, allows a direct comparison between children with idiopathic GHD and other GH-deficient patients. Because of the complex effects of puberty on growth, this model is only valid for prepubertal patients. The response to treatment is expressed as studentized residuals:

$$\frac{\text{Observed height velocity} - \text{predicted height velocity}}{\text{Standard error}}$$

Thus, a positive value indicates a response that is greater than that predicted from the data on idiopathic GH-deficient patients, and a negative value indicates a poorer response. Figure 1 illustrates the studentized residuals of the 25 prepubertal children in our cohort in whom all the relevant information was available, plotted against their predicted 1st-year height velocity. Of the 25 prepubertal children, 19 (76%) had a poorer response than predicted, and the higher the predicted height velocity, the greater the shortfall between the predicted and observed responses.

As a large proportion of the patients had entered puberty by the end of the 2nd year, the data derived from 20 children who remained prepubertal during the first 2 years of therapy were analyzed separately. Median age at the start of treatment was 8.7 years, and median height was similar to the cohort as a whole at –2.4 SDS. The response to GH is displayed in table 3. There is a significant increase in the height velocity in the 1st year of treatment, which is similar to the cohort as a whole, with an attenuated response in the 2nd year. The height SDS increased from –2.4 before treatment to –1.7 after 2 years of therapy, with no significant change in bone age delay.

Adverse Events

The 102 patients in this study received a total of 326 years of GH treatment. Only 40 adverse events were reported, and these consisted mainly of relatively minor events such as gastro-enteritis, allergies and upper respiratory tract disorders (table 4). In 14 patients some progression of the features of NF was noted, with 9 patients having an increase in the size of café-au-lait spots, 4 with new neurofibromas and 1 with an increase in the size of a prelumbar mass which was presumably a neurofibroma. Serious adverse events occurred in 5 patients, all of whom had previously had an intracranial tumour.

One girl had an optic glioma removed at the age of 3 years. She developed panhypopituitarism and was given GH (at a dose of 0.14 mg/kg/week), thyroxine and corticosteroid replacement therapy from the age of 3.8 years.

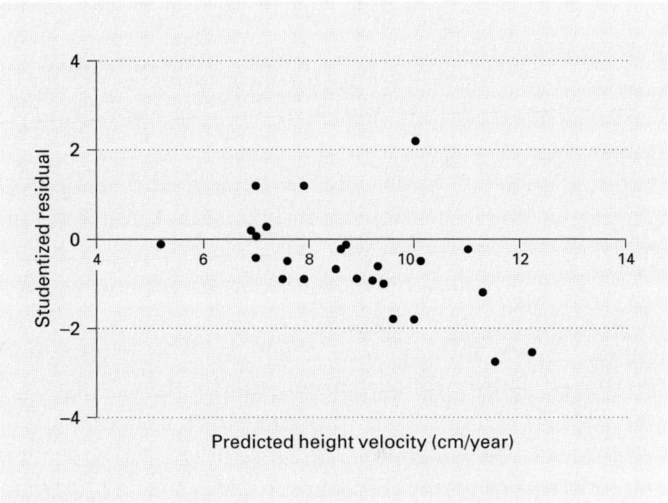

Fig. 1. Comparison of observed 1st-year height velocity of 25 prepubertal children with predicted height velocity based on a cohort of patients with idiopathic GHD. The studentized residual is the observed minus the predicted height velocity divided by its standard error.

Table 4. Adverse events over 3 years of GH treatment

Adverse event	Patients
Progression of NF	
Increasing size of café-au-lait spots	9
New neurofibromas	4
Increased size of prelumbar mass (?neurofibroma)	1
Recurrence of optic glioma	2
New intracranial tumour	
Right cerebral hemisphere astrocytoma	1
Pilocytic astrocytoma (?site)	1
Brainstem glioma	2
Other (gastro-enteritis, allergies, chest infection)	19

After 1 year on treatment a routine magnetic resonance imaging (MRI) scan revealed a recurrence of the optic glioma, and GH was stopped.

One girl was given radiotherapy at the age of 4 years for an optic glioma. At the age of 13.9 years she was started on GH at a dose of 0.14 mg/kg/week, and 6 months later a routine MRI scan revealed a pilocytic astrocytoma, though there was no information about the site of this lesion.

One boy had an optic nerve glioma diagnosed at the age of 5 years and was started on GH 4 years later. After 5.6 years of GH therapy at 0.21 mg/kg/week he developed headaches, ataxia and drowsiness, and a brainstem glioma was diagnosed.

One boy was irradiated for an optic glioma at the age of 6 years. He became GH deficient and was started on GH replacement therapy at the age of 13 years at a dose of 0.28 mg/kg/week. A year later he developed neurological symptoms and an MRI scan showed a brainstem tumour.

Another boy underwent cranial irradiation at the age of 5 years for an optic glioma. Deficiencies of TSH, gonadotrophin and GH were diagnosed 8 years later, and he started appropriate replacement therapy. After 2.3 years of GH at a dose of 0.15 mg/kg/week, a recurrence of the tumour was diagnosed, and he underwent a second course of radiotherapy. Five months later a follow-up scan revealed a right cerebral hemisphere astrocytoma, and GH was discontinued.

Discussion

This cohort of 102 patients represents the largest group of NF patients in whom the effects of GH treatment have been studied. As all the patients were GH deficient it is not surprising that this group differs significantly from other series of patients with NF that have previously been described. First, the median height of our patients was lower at –2.4 SDS, in keeping with the diagnosis of GHD. This compares with a median height at the 25th centile in an unselected NF population (equivalent to an SDS of –0.7) [1]. While the latter height is less of a deficit than that measured in our patients, it is still significantly different from that of an age-matched normal population. A recent study of 89 patients with NF [10] confirmed short stature in a significant proportion of patients (25.5% had height <10th centile), and, in addition, biochemical assessment demonstrated no underlying endocrinopathy, with normal GH status in all those tested. This suggests that there is a mild growth retardation associated with the disease itself, which is unrelated to GH status, and for which there is at present no explanation.

The prevalence of intracranial tumours in our cohort was 44% as compared with 15–20% in previous reports [3, 4, 11] on NF children. This is not surprising, as the treatment of these tumours with radiotherapy, and possibly the presence of the tumour itself, is associated with GHD, and, therefore, malignancy-associated cases are likely to be over-represented in our cohort. A study by Pierce et al. [6] of 24 patients with an optic glioma, half of whom had NF, found GHD in 15 out of 18 patients evaluated following treatment with radiotherapy. Precocious puberty occurred in 12% of our study population, all of whom had a tumour. Previous studies have found this figure to be approximately 3% in an unselected population of NF children [4, 5] and to occur almost exclusively in patients with intracranial tumours. When the excessive prevalence of tumours in our patients is taken into account, our observed incidence of precocious puberty is comparable to that reported in other studies.

Differences were also apparent in our cohort between patients with and without an intracranial tumour. The median pretreatment height SDS was significantly lower in the non-tumour group and remained so when the figures were adjusted for mid-parental height. This cannot be explained on the basis of weight or length at birth, severity of GHD as assessed by the results of standard GH stimulation tests or the duration of GHD. However, the incidence of precocious puberty was far greater in the tumour group (12/46) than in the non-tumour group (0/56), and this would seem the most likely explanation for the discrepancy in height between the groups. The fact that the pretreatment median bone age delay was less in the tumour group adds weight to this argument. However, this interpretation does not explain the difference between the two groups when only data from prepubertal patients were considered.

The response to GH therapy was similar in both groups of patients with a marked increase in height velocity in the 1st year associated with a period of catch-up growth. The height velocity was maintained well above baseline for the next 2 years, with a rate of growth comparable to that in an age-matched normal population. This resulted in a median gain in height SDS of 0.3 over 3 years of therapy. Although this appears to be only a modest gain in height, it is very likely that if left untreated the height SDS of this cohort would have continued to decrease over the 3 years, and therefore the real gain in height afforded by GH therapy is probably much greater.

Using the model developed in KIGS to predict 1st-year height velocity, it is clear that the majority of patients in our cohort had an initial response that was less than that observed in similar children with idiopathic GHD. The only factor known to influence the response to treatment that was not taken into account using this model is the peak GH response to standard GH stimulation tests, which correlates negatively with 1st-year height velocity. All of the idiopathic GHD patients, on whom the model is based, had a GH peak <10 µg/l, while only 86% of our cohort did. However, this difference is unlikely to have had a significant effect. Therefore, while children with NF and GHD benefit from GH therapy, their growth response is less impressive than that observed in patients with idiopathic GHD.

It has been suggested that one explanation for the modest response to therapy was the lower dose of GH used [12]. However, the median dose in our cohort is similar to that reported in groups of idiopathic GH-deficient patients taken from the KIGS [13, 14]. In addition, the model developed in KIGS, which predicts 1st-year height velocity, includes GH dose as one of the parameters from which the predicted growth response is calculated, and any differences in GH dose are, therefore, already taken into account. It is, therefore, clear that the poorer response to therapy in children with NF cannot be explained by the use of a lower GH dose. The most likely explanation for this observation is that in children with NF there is an additional adverse growth factor that is unrelated to endocrine abnormalities. It has long been recognized that short stature occurs in a significant proportion of children with NF [1], and recent data have shown that in the vast majority of cases this is associated with normal endocrine status [10]. Thus, even when GH replete, NF children will not grow as well as normal individuals, and it would, therefore, be predicted that following GH replacement the growth response in these children would be less than that observed in patients with idiopathic GHD. However, the question of GH dose may still be relevant. Supraphysiological doses of GH have been used with some success in Turner syndrome, another condition in which short stature is not a consequence of GHD. Thus, it is possible that the use of higher doses of GH would result in an improved growth response in GHD children with NF; however, the safety aspects would need careful consideration.

A total of 40 adverse events were reported in our cohort during GH therapy. Most of these were either relatively minor and unlikely to be directly attributable to GH therapy (gastro-enteritis, allergies, upper respiratory tract disorders) or were concerned with progression of features of NF. It has been well documented that the size of café-au-lait spots increases during childhood [15] to such an extent that the size criterion for the definition of a spot in prepubertal children is one third of that in adults (0.5 compared with 1.5 cm). Neurofibromas are also known to increase greatly both in size and number during puberty, being present in only 15% of children aged 6 years or younger in one study [16]. Therefore, the changes in café-au-lait spots and neurofibromas seen in 13% of our patients over a median follow-up period of 2.7 years do not represent an increase in disease progression above that which would normally be expected in a population of children with NF.

Serious adverse events occurred in 5 patients. All had a previous diagnosis of an optic glioma, for which 3 had received radiotherapy, and 1 had undergone surgery. Three of the patients developed a second intracranial tumour, 1 had a recurrence of the optic tract glioma, and 1 had a recurrence followed by a second intracranial tumour. Several authors have studied the rates of recurrence of optic gliomas and the incidence of second intracranial tumours.

Listernick et al. [4] reported tumour progression in 3 of 26 children with an optic glioma and NF who were followed up for a median of 2.4 years. A similar progression rate of 14% was observed in 14 children with NF who were reviewed for a median of 6 years following treatment for an optic glioma [17]. A review by Alvord and Lofton [18] of all the previous literature reported on a total of 623 patients with optic gliomas. A recurrence rate of approximately 30% by 10 years was calculated for 20 patients under the age of 20 years with NF who underwent surgery. A further 37 patients with NF who were treated with radiotherapy or simply observed had a progression/recurrence rate of 40% at 10 years.

Second intracranial tumours have been reported fairly frequently in children with optic gliomas and NF. Kuenzle et al. [19] described 21 such children, of whom 11 were found to have a cerebral tumour during a follow-up period of 9 years. Hochstrasser et al. [11] studied a group of 133 NF children, of whom 24 had intracranial tumours, of which 20 were optic gliomas. Five patients (4 with optic glioma) also had evidence of a second distinct intracranial tumour. In a review of NF patients that was not limited to children, Sørensen et al. [20] found a second malignancy (mostly intracranial) in 9 of 19 patients with a CNS tumour and 5 of 6 who had an optic glioma.

Our cohort included 43 children with intracranial tumours, of which 30 were optic gliomas. Progression of a treated optic glioma occurred in 2 of these 30 children (7%) during a median time of 2.7 years on GH therapy, a rate which is comparable to that previously reported. Four of the 43 patients with an intracranial tumour developed a second tumour during the follow-up period. More than half of the cohort described by Kuenzle et al. [19] developed a second tumour during 9 years of follow-up, and Hochstrasser et al. [11] and Sørensen et al. [20] found second tumours in 21 and 47%, respectively, of children with NF. Direct comparison of these studies with our cohort is difficult because of the differences in follow-up periods. However, even allowing for this, we do not feel that the rate of second-tumour development in our children was greater than that in comparable groups with NF. Furthermore, although the median period of GH therapy was only 2.7 years, the median time elapsed since the diagnosis of the first intracranial tumour was 7.4 years. Only the 2 patients who developed a brainstem glioma were symptomatic at the time of diagnosis of the second tumour. As there is no information available on whether any of the children were scanned prior to GH therapy, it is possible that some of the tumours were present before the start of treatment. Previously published data have also shown that malignancies do not occur more frequently in patients receiving GH replacement [21–23]. It, therefore, seems very unlikely that the use of GH affects the course of the various manifestations of NF and in particular the risk of tumour development.

Conclusions

Thus, we have shown that children with NF and GHD respond favourably to treatment with GH, though the growth response to therapy is not as good as that seen in patients with idiopathic GHD. We have also shown that GH therapy does not seriously influence the progression of any of the features of NF, including intracranial tumours, and is not associated with an excess of other adverse events. We can, therefore, conclude that, based on the data obtained from a substantial cohort of patients, GH treatment of NF patients with GHD is beneficial, in terms of improved growth rate, and is well tolerated.

References

1 Riccardi VM: Von Recklinghausen neurofibromatosis. N Engl J Med 1981;305:1617–1627.
2 Barker D, Wright E, Nguyen K, Cannon L, Fain P, Goldgar D, Bishop DT, Carey J, Baty B, Kivlin J, et al: Gene for von Recklinghausen neurofibromatosis is in the pericentromeric region of chromosome 17. Science 1987;236: 1100–1102.
3 Lewis RA, Gerson LP, Axelson KA, Riccardi VM, Whitford RP: Von Recklinghausen neurofibromatosis. II. Incidence of optic gliomata. Ophthalmology 1984;91:929–335.
4 Listernick R, Charrow J, Greenwald M, Mets M: Natural history of optic pathway tumors in children with neurofibromatosis type 1: A longitudinal study. J Pediatr 1994;125:63–66.
5 Habiby R, Silverman B, Listernick R, Charrow J: Precocious puberty in children with neurofibromatosis type 1. J Pediatr 1995;126:364–367.
6 Pierce SM, Barnes PD, Loeffler JS, McGinn C, Tarbell NJ: Definitive radiation therapy in the management of symptomatic patients with optic glioma. Survival and long-term effects. Cancer 1990;65:45–52.
7 Tanner JM, Whitehouse RH. Clinical longitudinal standards for height, weight, height velocity, weight velocity, and stages of puberty. Arch Dis Child 1976;51:170–179.
8 Niklasson A, Ericson A, Fryer JG, Karlberg J, Lawrence C, Karlberg P: An update of the Swedish reference standards for weight, length and head circumference at birth for given gestational age (1977–1981). Acta Paediatr Scand 1991;80:756–762.
9 Ranke M, Lindberg A, Guilbaud O: Prediction of growth in response to treatment with growth hormone; in Gunnarsson R, Ma R (ed): Progress in Growth Hormone Therapy – 5 Years of KIGS. Mannheim, J and J Verlag, 1994, pp 157–189.
10 Carmi D, Shohat M, Metzker A, Dickerman Z: Growth, puberty, and endocrine functions in patients with sporadic or familial neurofibromatosis type 1: A longitudinal study. Pediatrics 1999;103:1257–1262.
11 Hochstrasser H, Boltshauser E, Valavanis A: Brain tumors in children with von Recklinghausen neurofibromatosis. Neurofibromatosis 1988;1:233–239.
12 Saenger P: Growth hormone in von Recklinghausen's disease: Reckless or recommended? J Pediatr 1998;133:172–174.
13 Ranke MB, Price DA, Albertsson Wikland K, Maes M, Lindberg A: Factors determining pubertal growth and final height in growth hormone treatment of idiopathic growth hormone deficiency: Analysis of 195 Patients of the Kabi Pharmacia International Growth Study. Horm Res 1997;48:62–71.
14 Ranke MB, Guilbaud O: Growth response in prepubertal children with idiopathic growth hormone deficiency during the first two years of treatment with human growth hormone. Analysis of the Kabi Pharmacia International Growth Study. Acta Paediatr Scand Suppl 1991;379:109–116.
15 Listernick R, Charrow J: Neurofibromatosis type 1 in childhood. J Pediatr 1990;116:845–853.
16 Obringer AC, Meadows AT, Zackai EH: The diagnosis of neurofibromatosis-1 in the child under the age of 6 years. Am J Dis Child 1989; 143:717–719.
17 Janss AJ, Grundy R, Cnaan A, Savino PJ, Packer RJ, Zackai EH, Goldwein JW, Sutton LN, Radcliffe J, Molloy PT, et al: Optic pathway and hypothalamic/chiasmatic gliomas in children younger than age 5 years with a 6-year follow-up. Cancer 1995;75:1051–1059.
18 Alvord EC Jr, Lofton S: Gliomas of the optic nerve or chiasm. Outcome by patients' age, tumor site, and treatment. J Neurosurg 1988; 68:85–98.
19 Kuenzle C, Weissert M, Roulet E, Bode H, Schefer S, Huisman T, Landau K, Boltshauser E: Follow-up of optic pathway gliomas in children with neurofibromatosis type 1. Neuropediatrics 1994;25:295–300.
20 Sørensen SA, Mulvihill JJ, Nielsen A: Long-term follow-up of von Recklinghausen neurofibromatosis: Survival and malignant neoplasms. N Engl J Med 1986;314:1010–1015.
21 Ogilvy Stuart AL, Ryder WD, Gattamaneni HR, Clayton PE, Shalet SM: Growth hormone and tumour recurrence. BMJ 1992;304:1601–1605.
22 Blethen SL, Allen DB, Graves D, August G, Moshang T, Rosenfeld R: Safety of recombinant deoxyribonucleic acid-derived growth hormone: The National Cooperative Growth Study experience. J Clin Endocrinol Metab 1996;81:1704–1710.
23 Wilton P: Adverse events during growth hormone treatment: 5 years experience in the Kabi International Growth Study; in Gunnarsson R, Ma R (ed): Progress in Growth Hormone Therapy – 5 Years of KIGS. Mannheim, J and J Verlag, 1994, pp 291–307.

Hormone Research

Horm Res 2000;53(suppl 1):77–81

Growth Hormone Therapy in Noonan Syndrome

Christopher J.H. Kelnar

Department of Child Life and Health, University of Edinburgh, UK

Key Words

Noonan syndrome · Growth · Growth hormone · Cardiac features

Abstract

Noonan syndrome (NS) is an autosomal dominant disorder that can be difficult to diagnose. Growth retardation is a consistent feature, however, and although children are not typically growth hormone (GH) deficient, a minority may have suboptimal GH levels. An ongoing multicentre study examining the safety and efficacy of GH therapy in NS showed increases in height standard deviation scores (SDS; $p < 0.0001$) and height velocity ($p < 0.0001$) after 12 months of GH at pharmacological doses. There was no increase in mean maximal cardiac left ventricular wall thickness during the 12-month treatment period. Long-term follow-up data covering 3 years of GH therapy showed sustained increases in height SDS and height velocity compared with baseline ($p < 0.02$). Further large and appropriately controlled studies of high-quality design are essential to further our understanding of NS and to establish the long-term safety and efficacy of GH.

Introduction

In the 1930s, Ullrich and Turner described females with a syndrome of short stature, delayed sexual development and various dysmorphic features, most notably neck webbing (pterygium colli). It was subsequently demonstrated that this was caused by the absence or structural abnormality of the second X chromosome. However, male patients with a similar phenotype had also been observed as had some females with similar clinical features but a normal karyotype. In 1963 Noonan and Ehmke [1] described 6 male and 3 female patients with supravalvar pulmonary stenosis and a distinct clinical appearance. Characteristic dysmorphic features are now known to include abnormal facies with hypertelorism and a constellation of other features including mild mental retardation, cardiac defects, deafness, visual problems, hypogonadism with undescended testes, clotting disorders and short stature [2–7]. A comparison of the frequency of abnormalities between Turner and Noonan syndromes is summarized in table 1 [8].

Noonan syndrome (NS) is an autosomal dominant disorder with a reported incidence of between 1:1,000 and 1:2,500 livebirths with an equal sex ratio. However, problems with diagnosis arise because of the variability of the features in any given individual and the fact that the phenotype changes with age. Recent molecular genetic studies demonstrate linkage with chromosome 12 [9, 10], but at present, the diagnosis remains clinical with the typical facial features and at least two other features as key elements.

KARGER

Fax + 41 61 306 12 34
E-Mail karger@karger.ch
www.karger.com

0301–0163/00/0537–0077$17.50/0

Accessible online at:
www.karger.com/journals/hre

Dr. C.J.H. Kelnar
Department of Child Life and Health
University of Edinburgh, 20 Sylvan Place
Edinburgh EH9 1UW (UK)
Tel. +44 131 536 0807, Fax +44 131 536 0821, E-Mail dkcantavark@dial.pipex.com

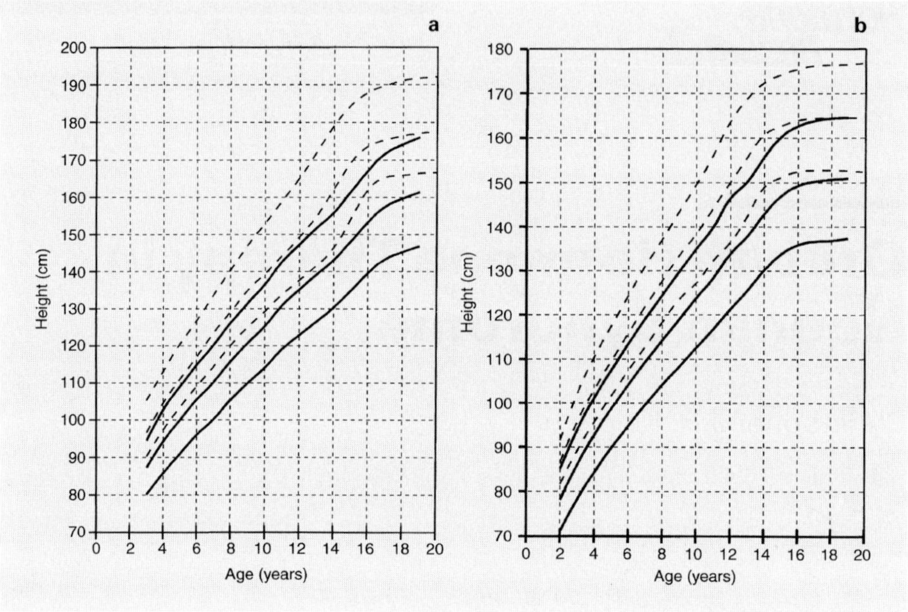

Fig. 1. Reference values for height in boys (**a**) and girls (**b**) with Noonan syndrome [from ref. 8].

Table 1. Comparison of the frequency of features between Noonan (n = 410) and Turner (n = 387) syndromes [from ref. 8]

Abnormal features	Noonan, %	Turner, %
Eyes	87	29
Ears	63	58
High-arched palate	51	61
Neck/pterygium colli	41	67
Low-set hairline	61	80
Thorax	64	76
Cubitus valgus	67	50
Skin	32	72
Cardiac defect (all)	56	20
Pulmonary stenosis	42	rare
Coarctation	3	14
Renal	25	57
Gonads		
Male	72	–
Female	rare	95
Mental handicap	44	11

Growth and Puberty

Growth retardation is a consistent feature of NS. The mean adult height is approximately 162.5 cm (males) and 153 cm (females) which is approximately the third centile for the normal population [5, 11, 12]. The mean growth curves for children with NS follow the third percentile through childhood, but then there is a fall below the 'normal' range as a result of an average 2-year delay in the onset of puberty which is typical in these individuals [11, 12]. Typically, final height is not achieved until the end of the second decade. Reference standards are shown schematically in figure 1 [8]. However, standards are based on relatively small numbers of subjects and largely cross-sectional data. There is a lack of prospective, high-quality longitudinal or mixed cross-sectional growth data for children with NS who have not been treated with growth promoting therapy which has implications for identifying and quantifying any treatment effects. In addition, there may be a bias in reporting towards shorter children with NS and those with more obvious dysmorphic features.

The Growth Hormone Axis

Children with NS are typically not growth hormone GH deficient, but there have been reports of impairment of the GH-insulin-like growth factor (IGF) axis [13–16]. A minority (10–45%) of children with NS have suboptimal [<20 mU/l (10 mg/l)] GH levels in response to standard GH stimulation tests [17–19]. Thus, levels of markers of GH action – IGF-1 and IGF-binding protein 3 are characteristically low and rise with GH therapy [14, 17]. There is no evidence for GH resistance in NS. It is unclear to what extent the 'low' to 'normal' stimulated GH levels simply reflect a spectrum of GH secretory ability among children with NS destined to be at the upper or lower ends of the

Table 2. Studies of GH therapy in NS

Study	n	Age years	GH dose IU/kg/week	Pre-treatment height SDS[a]	1-year height SDS[a]	3-year height SDS[a]	Pretreatment height velocity cm/year	1-year height velocity cm/year	3-year height velocity cm/year	Final height data Δheight SDS
Ahmed et al. [14]	6	11.2	0.6	−2.85	−2.56	ND	4.8	7.4	ND	ND
Otten [19]	55	10.6	0.61	−3.1	−2.6	−2.4	4.3	7.0	ND	ND
Romano et al. [18]	42	ND	0.62	−3.3	−2.8	−2.4	4.3	8.0	ND	ND
Thomas and Stanhope [28]	5	3.9	1.01	−3.3	−2.4	ND	ND	ND	ND	ND
De Schepper et al. [29]	23	10.3	1.05	−2.28	−1.78	ND	4.5	8.5	ND	ND
Municchi et al. [30]	4	13.5	0.5	ND	ND	ND	ND	ND	ND	−1.2 to +1.8
Cotterill et al. [17]/ Macfarlane et al. [32]	30	8.9	1.19	−3.01[b]	−2.36[b]	−1.9	4.9	8.9	5.8	ND

ND = No data available/reported.
[a] Tanner Whitehouse; [b] UK 1990 growth standards [32].

NS height spectrum (as with the spectrum of GH secretory ability seen in the normal population) [20] or whether they represent a specific NS-related problem.

GH Therapy

Possible reasons for wishing to treat NS patients with GH include significant short stature compared to normal peers, the superficial analogy to Turner syndrome, possible impairment of the GH-IGF axis and the commonness of the condition. Should patients with, generally, no recognizable abnormalities in the GH-IGF-I axis be given daily GH treatment to promote growth? If so, what is the optimum treatment regimen, how can we predict who will respond, and what are the most appropriate end points by which to measure the efficacy of treatment?

An important added consideration in deciding on the use of GH therapy to enhance growth in children with NS concerns the nature of the cardiac defects and (more than just theoretical) concerns about a deleterious effect of GH on their progression. Characteristic cardiac anomalies in NS are pulmonary valve stenosis (62%), unexplained left ventricular hypertrophy (20%) and secundum atrial septal defect [7, 21–26]. Patients with a mean maximal left ventricular wall thickness of >1 cm or other echocardiographic features of hypertrophic cardiomyopathy (which is not usually apparent clinically) [17, 23] may be at particular risk because of the effect of GH on cardiac muscle mass in other situations [27].

Seven major studies [14, 17–19, 28–30] have reported on the use of GH therapy in NS. They have now been summarized in detail [31], and some of the findings are shown in table 2. However, these studies suffer collectively and individually from a number of deficiencies: (1) they are observational studies with no randomization or placebo or other control groups; (2) there is inconsistent validation of the diagnosis (similar conditions include Leopard and cardio-facio-cutaneous syndromes), particularly in the large(r) multicentre studies that use existing database records, where it is likely that some patients with other conditions are included in the analyses; (3) because of differing views about the nature of the problem (if any) with the GH-IGF-1 axis in NS, a wide GH dose range has been employed ranging from physiological replacement [~14 U/m^2/week (~0.7 mg/m^2/day)] to pharmacological treatment [28 U/m^2/week (~1.3 mg/m^2/day)], comparable with treatment in Turner syndrome]; (4) most are short-term studies, with only one of the studies following to final height in only 4 individuals; (5) surrogate markers (e.g., predicted adult height and target height) are used for efficacy/final height prediction, and these are notoriously unreliable – in particular, comparisons are made with published reference NS charts (which are largely cross-sectional, derived from relatively small numbers and subject to reporting bias), predicted adult height (particularly problematic in children with growth abnormalities), parental target height (similarly problematic) and pretreatment height velocities which are useful for comparing short-term treatment effects only; (6) the confounding effects of puberty on growth velocity are poorly accounted for; (7) in general, there are no properly randomized control groups – there is a tendency to use as comparison groups those who refuse treatment or who cannot be included for other reasons (e.g., ethical, social, cardiac factors); (8) there is likely to be ascertainment bias in individuals recruited (most dysmorphic/shortest); (9) there is seldom ascertainment of compliance and no systematic intention-to-treat analysis of 'drop outs', and (10) few studies look systematically at cardiac and other potential side-effects.

An ongoing UK multicentre study [17, 32] is examining the safety and efficacy of pharmacological GH therapy [28 U/m^2/week (~ 1.3 mg/m^2/day)]. Entry criteria for this study included: diagnosis of NS confirmed by a single observer (a clinical geneticist with a particular interest in and experience of the condition); height standard deviation score (SDS) less than or equal to –2 [33]; prepubertal, and no evidence of cardiomyopathy (normal maximal left ventricular wall thickness (i.e. <1 cm) by two-dimensional echocardiography. Thirty children (19 boys, 11 girls) aged 8.9 ± 0.5 (range 4.8–13.7) were recruited. A comparison group (n = 10; 4 girls and 6 boys; mean age 8.8 ± 0.7, range 6.3–11.8 years) was also studied. These children were not recruited to the study group because consent was not given (n = 2), or because entry criteria were not met [height SDS > –2.0 (n = 7), or evidence of mild hypertrophic cardiomyopathy (n = 1)].

In the treatment group, 27 of the 30 children completed 12 months of GH therapy. Height SDS increased from –3.01 ± 0.10 to –2.36 ± 0.10 ($p < 0.0001$) after 12 months, and height velocity increased from 4.9 ± 0.2 to 8.9 ± 0.3 cm/year at 6 months and was 8.1 ± 0.4 cm/year ($p < 0.0001$) between 6 and 12 months. Height velocity SDS increased from –0.7 ± 0.15 to +2.42 ± 0.32 over 12 months ($p < 0.0001$). The increase in height velocity was more than 2 cm/year in 24 patients. IGF-I increased from 121 ± 13 to 240 ± 22 (μg/l at 12 months ($p < 0.0001$). IGF-binding protein 3 increased from 2.65 ± 0.20 to 4.01 ± 0.42 mg/l at 12 months ($p = 0.0009$). In the comparison group there was no change in height SDS (–2.03 ± 0.19), height velocity (4.4 ± 0.24 cm/year) or height velocity SDS (–1.08 ± 0.21).

There was no increase in mean maximal cardiac left ventricular wall thickness during the study in either the treatment group (12-month values 0.63 ± 0.02 cm at the mitral valve level and 0.66 ± 0.02 cm at the papillary muscle level) or in the comparison group (0.63 ± 0.04 cm at the mitral valve level and 0.61 ± 0.03 cm at the papillary muscle level).

In this cohort of 30 children starting GH therapy, 3 did not complete 12 months of therapy, 3 showed no improvement in growth (1 due to non-compliance), and 1 patient was withdrawn on medical grounds (mitral regurgitation progressing to cardiac failure). Thus, GH was effective in 23 of the treated patients, with significant increases in height SDS and height velocity over 12 months of therapy. Importantly, any theoretical abnormal anabolic effects of GH on myocardial thickness were not confirmed, and no patient developed features of hypertrophic cardiomyopathy.

Longer-term follow-up data are now available [32]. In the treatment group (n = 23 at 3 years from start of GH therapy), height SDS improved from –2.7 ± 0.4 at start of therapy to –1.9 ± 0.9 after 3 years ($p < 0.001$). Height velocity increased from 4.4 ± 1.7 cm/year in the year prior to treatment to 8.5 ± 1.6 ($p < 0.001$), 6.2 ± 1.7 ($p < 0.001$) and 5.8 ± 1.7 cm/year ($p < 0.02$ compared with baseline) during the 1st, 2nd and 3rd years of GH therapy, respectively. During the 1st year of therapy, 18 patients (78.3%) showed an increase in height velocity of at least 2 cm/year. During the 2nd and 3rd years, this fell to 52.2 and 30.4% of the patients, respectively. Over 3 years, the control group showed a non-significant improvement in height SDS (from –2.7 ± 0.6 to –2.3 ± 0.7, $p = 0.21$). Mean height velocity over 3 years was 5.5 ± 0.8 cm/year in the control group versus 6.9 ± 1.2 cm/year in the treatment group.

Conclusions

From these and the other (albeit imperfect) studies [14, 17–19, 28–30, 32] some conclusions can be drawn. Current evidence for the use of GH in NS remains weak, though it appears effective in increasing short-term growth and is well tolerated. Nevertheless, both short- and longer-term studies have demonstrated a consistent and significant increase in height velocity in children with NS treated with GH. There appears to be a dose-related response with greater efficacy from higher (pharmacological) doses of GH rather than with physiological replacement. No adverse effects on bone maturation have been observed. Therapy has failed in some patients, however, and likely reasons include poor compliance, the heterogeneity of the condition and the difficulties with being certain about the diagnosis in some individuals in the absence of a molecular genetic marker for the condition. It is not currently possible to predict poor responders in advance.

In the future, investigation of the genetic, metabolic and auxological variables that affect responses to treatment and incorporation of these into prediction models for an individual with NS will be helpful, particularly when the diagnosis can be confirmed by molecular genetic techniques. This should allow treatment decisions to be based on a realistic expectation of benefits and should aid the design of individual treatment regimens and identification of non-responding or non-compliant patients during treatment. Prediction models can also be used as a tool to further our understanding of the pathophysiology of growth disorders by the identification of disparate types of response to therapy in what had previously been considered a group with a single diagnostic label.

The long-term safety of GH therapy should be considered in treatment decision-making; current knowledge suggests that it is important to exclude the presence of hypertrophic cardiomyopathy on echocardiography before considering GH therapy, and patients on treatment should be monitored annually to check for its development.

Incontrovertible evidence for the efficacy and safety of GH therapy in NS will come only from appropriately controlled studies of sufficiently larger numbers of subjects who have been defined on a molecular genetic basis and followed to final height. This will require international co-operation.

References

1 Noonan J, Ehmke DA: Associated non-cardiac malformations in children with congenital heart disease. J Pediatr 1963;63:468.
2 Mendez HMM, Opitz JM: Noonan syndrome: A review. Am J Med Genet 1985;21:493–506.
3 Allanson JE, Hall JG, Hughes HE, Preus M, Witt RD: Noonan syndrome. A changing phenotype. Am J Med Genet 1985;21:507–514.
4 Allanson JE: Noonan syndrome. J Med Genet 1987;24:9–13.
5 Sharland M, Burch M, McKenna WJ, Patton MA: A clinical study of Noonan syndrome. Arch Dis Child 1992;67:178–183.
6 Sharland M, Taylor R, Patton MA, Jeffery S: Absence of linkage of Noonan syndrome to the neurofibromatosis type 1 locus. J Med Genet 1992;29:188–190.
7 Caralis DG, Char F, Graber JD, Voigt GC: Delineation of multiple cardiac anomalies associated with Noonan syndrome in an adult and review of the literature. Johns Hopkins Med J 1974;134:346–353.
8 Ranke MB: Turner and Noonan syndromes; in Kelnar CJH, Savage MO, Stirling HF, Saenger P (eds): Growth Disorders – Pathophysiology and Treatment. London, Chapman & Hall, 1998, pp 623–639.
9 Jamieson CR, van der Burgt I, Brady AF, van Reen M, Elsawi MM, Hol F, Jeffery S, Patton MA, Mariman E: Mapping a gene for Noonan syndrome to the long arm of chromosome 12. Nat Genet 1994;8:357–360.
10 van der Burgt I, Berends E, Lommen E, van Beersum S, Hamel B, Mariman E: Clinical and molecular studies in a large Dutch family with Noonan syndrome. Am J Med Genet 1994;53: 187–191.
11 Witt DR, Keena BA, Hall JG, Allanson JE: Growth curves for height in Noonan syndrome. Clin Genet 1986;30:150–153.
12 Ranke M, Heidemann P, Knupfer C, Enders H, Schmaltz AA, Bierich JR: Noonan syndrome: Growth and clinical manifestations in 144 cases. Eur J Pediatr 1988;148:220–227.
13 Spadoni GL, Bernardini S, Cianfarani S: Spontaneous growth hormone secretion in Noonan syndrome. Acta Paediatr Scand 1990(suppl 367):157.
14 Ahmed ML, Foot AB, Edge JA, Lamkin VA, Savage MO, Dunger DB: Noonan's syndrome: Abnormalities of the growth hormone/IGF-I axis and the response to treatment with human biosynthetic growth hormone. Acta Paediatr Scand 1991;80:446–450.
15 Elders MJ, Char F: Possible etiologic mechanisms of the short stature in the Noonan syndrome. Birth Defects Orig Artic Ser 1976;12: 127–133.
16 Tanaka K, Sato A, Naito T, Kuramochi K, Itabashi H, Takemura Y: Noonan syndrome presenting with growth hormone neurosecretory dysfunction. Intern Med 1992;31:908–911.
17 Cotterill AM, McKenna WJ, Brady AF, Sharland M, Elsawi M, Yamada M, Camacho-Hubner C, Kelnar CJ, Dunger DB, Patton MA, Savage MO: The short-term effects of growth hormone therapy on height velocity and cardiac ventricular wall thickness in children with Noonan's syndrome. J Clin Endocrinol Metab 1996;81:2291–2297.
18 Romano AA, Blethen SL, Dana K, Noto RA: Growth hormone treatment in Noonan syndrome: The National Cooperative Growth Study experience. J Pediatr 1996;128(suppl): 18–21.
19 Otten BJ: Short stature in Noonan syndrome: Demography and response to growth hormone treatment in the Kabi International Growth Study; in Ranke MB, Gunnarsson R (eds): Progress in Growth Hormone Therapy – 5 Years of KIGS. Mannheim, J und J Verlag, 1996, pp 206–215.
20 Albertsson-Wikland K, Rosberg S, Isaksson O, Westphal O: Secretory pattern of growth hormone in children of differing growth rates. Acta Endocrinol (Copenh) 1983;103(suppl 256):72.
21 Burch M, Sharland M, Shinebourne E, Smith G, Patton MA, McKenna W: Cardiologic abnormalities in Noonan syndrome: Phenotypic diagnosis and echocardiographic assessment of 118 patients. J Am Coll Cardiol 1993;22:1189–1192.
22 Wilmshurst PT, Katritsis D: Restrictive and hypertrophic cardiomyopathies in Noonan syndrome: The overlap syndromes. Heart 1996;75:94–97.
23 Burch M, Mann JM, Sharland M, Shinebourne EA, Patton MA, McKenna WJ: Myocardial disarray in Noonan syndrome. Br Heart J 1992;68:586–588.
24 Cooke RA, Chambers JB, Curry PV: Noonan's cardiomyopathy: A non-hypertrophic variant. Br Heart J 1994;71:561–565.
25 Nishikawa T, Ishiyama S, Shimojo T, Takeda K, Kasajima T, Momma K: Hypertrophic cardiomyopathy in Noonan syndrome. Acta Paediatr Jpn 1996;38:91–98.
26 Ishizawa A, Oho S, Dodo H, Katori T, Homma S: Cardiovascular abnormalities in Noonan syndrome: The clinical findings and treatments. Acta Paediatr Jpn 1996;38:84–90.
27 Osterziel KJ, Strohm O, Schuler J, Friedrich M, Hanlein D, Willenbrock R, Anker SD, Poole-Wilson PA, Ranke MB, Dietz R: Randomised, double-blind, placebo-controlled trial of human recombinant growth hormone in patients with chronic heart failure due to dilated cardiomyopathy. Lancet 1998;351:1233–1237.
28 Thomas BC, Stanhope R: Long-term treatment with growth hormone in Noonan's syndrome. Acta Paediatr 1993;82:853–855.
29 De Schepper J, Otten BJ, Francois I, Bourguignon J-P, Craen M, van der Burgt I, Massa GC: Growth hormone therapy in pre-pubertal children with Noonan syndrome – first year growth response and comparison with Turner syndrome. Acta Paediatr 1997;86:943–946.
30 Municchi G, Pasquino AM, Pucarelli I, Cianfarani S, Passeri F: Growth hormone treatment in Noonan syndrome: Report of four cases who reached final height. Horm Res 1995;44:164–167.
31 Cotterill AM: Effectiveness of growth hormone therapy in Noonan syndrome; in Hindmarsh PC (ed): Current Indications for Growth Hormone Therapy. Basel, Karger, 1999, pp 118–127.
32 Macfarlane CE, Brown DC, Brady AF, Johnston LB, Patton MA, Dunger DB, Savage MO, Kelnar CJH: Growth hormone therapy and growth in children with Noonan's syndrome: Results of three years follow-up (abstract 373). Horm Res 1997;48(suppl 2):72.
33 Freeman JV, Cole TJ, Chinn S, Jones PRM, White EM, Preece MA: Cross-sectional stature and weight reference curves for the UK, 1990. Arch Dis Child 1995;73:17–24.

Hormone Research

Horm Res 2000;53(suppl 1):82–86

Treatment of Juvenile Rheumatoid Arthritis with Growth Hormone

Dominique Simon[a] Anne-Marie Prieur[b] Paul Czernichow[a]

[a]Service d'Endocrinologie et Diabétologie Pédiatriques, Hôpital Robert-Debré, et [b]Service d'Immuno-Hématologie, Hôpital Necker-Enfants Malades, Paris, France

Key Words

Juvenile arthritis · Growth hormone · Bone metabolism · Growth

Abstract

Severe growth retardation and profoundly altered body composition are observed in children with juvenile chronic arthritis receiving glucocorticoids. This study assessed the effects of growth hormone (GH) on height velocity, body composition and bone density. Fourteen patients were treated with GH (1.4 U/kg/week) for 1 year and then studied for a 2nd year off GH. The treatment increased insulin-like growth factor 1 and insulin-like growth factor binding protein 3 plasma levels. All patients showed an increase in height velocity. Lean body mass increased by 12%. After the cessation of GH therapy, height velocity fell to pretreatment values, and weight and fat mass increased markedly. Bone formation and resorption markers significantly increased during treatment and returned to pretreatment values after discontinuation of GH treatment. These results suggest that GH may partially counteract the adverse effects of glucocorticoids on growth and metabolism in patients with chronic inflammatory disease.

Introduction

Growth retardation is a frequent complication of juvenile chronic arthritis (JCA) in children. In systemic forms of the disease, severe impairment of height velocity and metabolic abnormalities are induced by long-term glucocorticoid (GC) therapy which is frequently necessary. We assessed the effects of a 1-year course of recombinant human growth hormone (GH) on growth, body composition, bone metabolism markers and bone density in 14 children with JCA receiving GC therapy. We also evaluated the potential adverse effects of GH, particularly with regard to glucose tolerance.

Patients and Methods

Inclusion Criteria

Inclusion criteria were: age over 3 years; prepubertal status; bone age (BA) under 12 years for boys and under 10 years for girls; height less than –2 standard deviation scores (SDS) for chronological age (CA); height velocity less than – 1 SDS for CA during the year preceding inclusion, and GC treatment for more than 2 years with a stable dose for more than 6 months prior to inclusion.

Patient Characteristics

A group of 14 children with severe systemic and/or polyarticular JCA were enrolled. All children were receiving prednisone. Despite a mean daily prednisone dose of 0.38 mg/kg/day, the inflammatory process was persistent, with a mean erythrocyte sedimentation rate (ESR) at enrollment of 58 mm at 1 hour.

KARGER

Fax + 41 61 306 12 34
E-Mail karger@karger.ch
www.karger.com

0301–0163/00/0537–0082$17.50/0

Accessible online at:
www.karger.com/journals/hre

Prof. P. Czernichow
Service d'Endocrinologie et Diabétologie Pédiatriques
Hôpital Robert-Debré, 48, bd Sérurier
F–75019 Paris (France)
Tel. +33 1 400 32360, Fax +33 1 400 32429

Study Design

Patients received GH (Genotonorm®; Pharmacia & Upjohn) at a dose of 1.4 IU/kg/week (0.42 mg/kg/week) given by daily subcutaneous injection. GH was administered for 1 year, and the patients were monitored for a 2nd year after the cessation of treatment. Height, weight and height velocity were compared with reference values for the French population [1]. Puberty was graded with Tanner scores. Joint involvement was assessed clinically by the same rheumatologist. BA was assessed according to Greulich and Pyle [2].

EST, the A1c fraction of glycosylated haemoglobin (HbA_{1c}), insulin-like growth factor (IGF) 1 and IGF-binding protein 3 (IGFBP-3) were measured at baseline and after 1, 6, 12 and 24 months. An oral glucose tolerance test was also performed at these intervals.

The study was approved by the Paris-Saint Louis Ethics Committee. Informed parental and patient consent was obtained in every case.

Methods of Assessment

Body composition and bone density were measured by means of dual-energy X-ray absorptiometry (Hologic QDR1000W/892 mef 1990) [3]. GH secretion was evaluated by means of the glucagon-betaxolol stimulation test and spontaneous nocturnal secretion. GH serum concentrations were measured by radio-immunoassay (normal values as the GH peak after stimulation >10 μg/l; nocturnal GH as the area under the curve >65 μg/l/8 h). Serum IGF-1 was assayed after separation from IGFBP by chromatography in acid conditions. IGF-1 content was measured by radio-immunoassay (normal values for CA 50–290 μg/l). IGFBP-3 was measured directly in serum by using a commercial radio-immunoassay kit (normal values for CA 2.20–6.4 mg/l).

Blood levels of calcium, phosphora, albumin and alkaline phosphatases and urinary calcium were measured by standard techniques. The osteocalcin concentration was measured by a solid-phase 'sandwich' immunoradiometric assay (ELSA-OSTEO; CIS Bio International). Procollagen (PICP. I) was measured by a competition radio-immunoassay technique (Procollagen PICP, Orion Diagnostics).

The total amounts of urinary deoxypyridinoline and pyridinoline were measured in hydrolyzed samples using fluorometric detection.

The oral glucose tolerance test (glucose concentration 2 g/kg, with measurements of plasma glucose and serum insulin levels at 0, 60 and 120 min) was performed after an overnight fast. Insulin levels were measured by radio-immunoassay. HbA1c was measured by high-performance liquid chromatography (normal values: 5.1 ± 0.6%).

Statistical Analysis

Values before, during and after treatment were compared using Student's paired t test.

Results

Patient Characteristics at Enrollment

The mean age at the onset of GH treatment was 9 years and 8 months. All the patients were extremely short and had a very low height velocity. Their body weight (mean –1.9 SDS for CA) showed a smaller deficit, reflecting the weight excess of these children on long-term GC treatment. The mean bone age delay for CA was 2 years and 4 months. Stimulated GH secretion was normal except in 1 patient. Spontaneous nocturnal secretion was below –2 SDS in 7 of the 14 patients. The values for IGF-1 were normal in 12 and low in 2 of the 14 patients. IGFBP-3 plasma concentrations were normal in all patients.

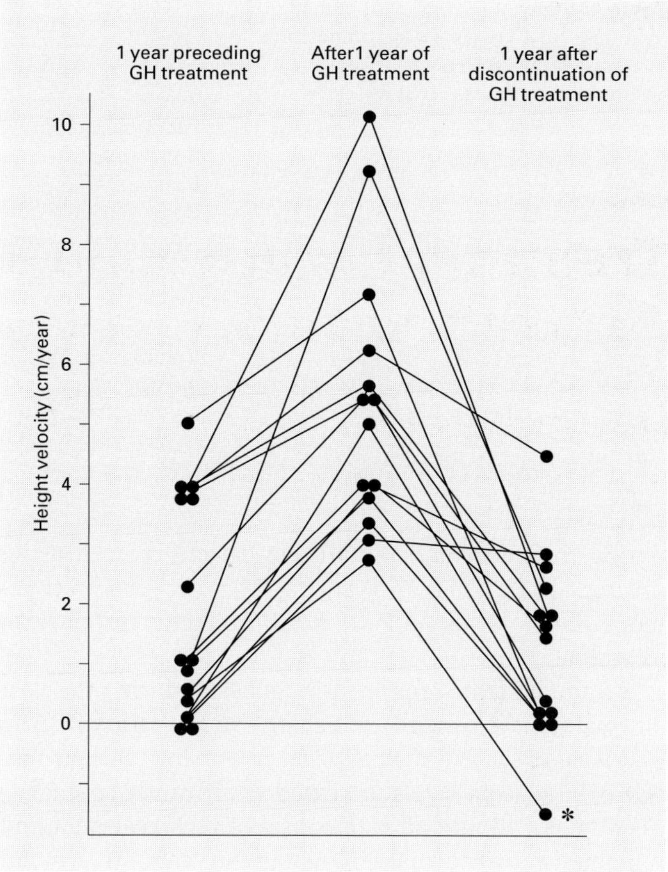

Fig. 1. Effect of GH treatment on height velocity in 14 patients with JCA receiving long-term steroid treatment. Asterisk: the negative height velocity observed during the 2nd year of follow-up in 1 patient is due to vertebral compression.

Effects of GH Treatment on Growth Parameters

During the 1-year treatment period, the mean height velocity increased from 1.9 to 5.4 cm/year ($p = 0.0001$). The height velocity increased in all patients, but this increase varied markedly between patients. Three patients had partial catch-up growth, and 6 patients had a height velocity below –2 SDS for CA. The height velocity returned to pretreatment values when GH treatment was stopped (fig. 1). There was a highly significant inverse correlation ($r = -0.7$; $p = 0.003$) between the height velocity

Table 1. Effect of recombinant human GH treatment for 1 year on body composition as assessed by dual-energy X-ray absorptiometry (mean ± SD)[a]

	Months of study (n = 14)			
	0	6	12	24
Fat mass, kg	5.5 ± 2.4	4.2 ± 1.7**	4.4 ± 1.8**	8.2 ± 4.0**
Body weight, %	23.7 ± 3.4	18.2 ± 2.4*	18.4 ± 2.5*	27.1 ± 5.8**
Lean mass, kg	16.4 ± 5.2	17.6 ± 5.9**	18.4 ± 6.5**	19.8 ± 6.9**
Body weight, %	72.1 ± 3.4	76.8 ± 2.3**	76.4 ± 2.2**	68.6 ± 4.7**

Normal values for prepubertal, normoponderal children: lean mass, 79.6 ± 2.3% of body weight; fat mass, 18.1 ± 0.8% of body weight.

[a] Values after 6 and 12 months of treatment were compared with values at enrollment. Values after 24 months (off treatment) were compared with values after 12 months (on treatment).

* $p < 0.05$; ** $p < 0.01$ (Student's paired t test).

during treatment and prednisone dosage, as well as an inverse correlation ($r = -0.56$; $p > 0.05$) between pretreatment IGF-1 levels and the height velocity response to treatment.

All the patients remained prepubertal during the year on treatment. No significant change in the ratio of BA to CA was observed during GH treatment. IGF-1 and IGFBP-3 plasma levels increased significantly ($p < 0.01$) as early as 1 month after the onset of GH treatment and remained above normal throughout the year of treatment. When GH was stopped, IGF-1 and IGFBP-3 levels returned to pretreatment values.

Effects of GH Treatment on Nutrition and Metabolism

Most of the children had a low caloric intake for their age (mean 1,305 kcal/day) which did not increase significantly during the year of treatment.

The mean lean mass increased by 12%, and the mean fat mass fell by 20% during treatment (table 1). After the cessation of treatment, the fat mass again increased markedly (+87%). The total lean mass continued to increase (+7%), but less than during the year on treatment; thus, the total percentage of lean mass diminished.

Effect on Bone Metabolism

All patients at inclusion showed severe bone demineralization (mean bone density –3.7 SDS for CA). Routine bone metabolism markers were normal. Compared with pretreatment values, bone formation markers (osteocalcin and C-terminal propeptide of type 1 procollagen blood levels) and bone resorption markers (urinary hydroxyproline, pyridinoline and deoxypyridinoline levels) significantly increased during treatment ($p < 0.01$) and returned to pretreatment values after discontinuation of GH treatment. However, bone density was not changed during GH administration despite active modification of bone turnover as shown by the variations of biochemical markers.

Effect of GH Treatment on Glucose Tolerance

Compared with baseline values, glycaemia during GH treatment was slightly increased in the fasting state and after a glucose load, but the differences were not significant. Increased serum insulin levels were also observed in all 14 patients, both in the fasting state and after an oral glucose load. HbA_{1c} increased significantly to above our upper limit of normal in 8 patients after the year on treatment. After the cessation of treatment, blood HbA_{1c} and fasting insulin levels returned to pretreatment values. Serum insulin levels after an oral glucose load remained insignificantly higher than before treatment.

Other Effects of GH Treatment

GH had no clinical impact on the arthritis. The absence of any significant change in inflammatory activity is reflected by the stability of GC doses and the stability of the ESR. Three children complained of muscle pain during the first few weeks of GH treatment, but this disappeared spontaneously. No worsening of scoliosis was noted. One patient had osteochondritis of the hip, but this complication is frequent in JCA, and we thus decided to continue GH treatment; the lesions consolidated.

Discussion

The reasons for growth retardation in JCA are multifactorial, including the chronic inflammatory state, bone and joint lesions and undernutrition. The severe complications of active systemic JCA in terms of growth and metabolism are exacerbated by GH therapy. The mechanism of linear growth inhibition by GC has not been defined. The findings regarding the effects of GC on GH secretion are controversial [4–8]. It appears that GC may affect not so much the level of GH secretion, but rather the nyctohemeral GH secretion profile, resulting in delayed and attenuated nocturnal GH secretion peaks [9]. Butenandt [10] described reduced secretion of GH in children with JCA treated with GC. Like other authors [11–13], we have found a normal stimulated GH secretion peak. However, an attenuation of nocturnal GH peaks was found in half of our patients which may contribute to growth retardation.

Few studies have been reported on GH treatment in patients with JCA [10, 13–17]. An increase in height velocity was observed in most patients in these studies. However, the heterogeneity of the response to GH treatment in terms of height velocity in our patients must be emphasized. As growth rates returned to the normal range but were not markedly above normal, GH treatment prevented further height SDS deterioration, but did not produce catch-up growth. However, stabilization of the children's relative position on the growth curve indicates a benefit of GH treatment. The contribution of JCA disease activity to changes in growth is difficult to quantify. None of our patients had spontaneous remissions from the disease during the year of treatment, and their chronic inflammatory disease may be considered stable, as reflected by the stable doses of steroids and a stable mean ESR.

The actions of GH may be divided into direct effects and those that occur after the modulation of mediators such as IGF and IGFBP. Low IGF-1 and low IGFBP-3 levels have been found in catabolic states [18], and a substantial increase in serum IGF-1 concentrations may reverse catabolism [19]. Normal levels of somatomedins have been described in patients with Cushing syndrome and those treated with GC [20]. However, such patients may have a reduction in the biological activity of somatomedin [20, 21]. The negative correlation between pretreatment IGF-1 levels and height velocity in response to treatment demonstrates the importance of anomalies of the GH-IGF-1 axis in understanding abnormal growth in these patients. Before GH treatment, the levels of IGF-1 in our patients were at the low end of the normal range, possibly reflecting resistance to GH. The improvement in height velocity was associated with an increase in IGF-1 and IGFBP-3 plasma levels to above normal values, as well as with an increase in the ratio of IGF-1 to IGFBP-3, demonstrating increased free IGF-1 and, therefore, IGF-1 bioavailability.

The lipolytic effect of GH is well known [22, 23]. Our results show that the administration of GH may lead to a drastic reduction in the excess of fat mass induced jointly by JCA and GC treatment. These modifications reduced the cushingoid appearance of the children and provided an important psychological benefit [24].

In our study, we did not observe a significant change in bone density after 12 months of treatment. This may be linked with the limit of present technology to evaluate bone turnover which may be not precise enough for us to determine the effects induced by GH in such complex situations. However, in childhood-onset GH deficiency and osteopenia, bone mineral density increases after only a delay of 6–30 months of GH treatment [25–27]. Longer studies are now required to determine whether GH may a least partially improve bone density in children affected with severe systemic forms of JCA.

Numerous studies have shown the ability of GH to impair insulin sensitivity [28]. Diabetes mellitus has occasionally been reported in patients treated with GH [29]. In Cushing syndrome, the additional effects of excess endogenous GH may induce resistance to insulin [30]. In patients who have the risks of both exogenous GC and GH therapies, we and others [12] have observed a small but significant increase in blood glucose values and a more marked rise in blood insulin values, but no diabetes. The risk of diabetes and the consequences of long-term hyperinsulinism should be evaluated in long-term studies.

Conclusions

Overall, our data confirm that GH therapy may counteract the harmful effects of GC on growth and metabolism in patients with systemic JCA. Larger and longer controlled studies are needed to determine the long-term risks and to confirm the benefits of GH therapy for GC-dependent children. After completion of this study, the 14 patients were invited to participate in a new 3-year GH therapeutic trial, and all have agreed.

Acknowledgements

This study was supported by Pharmacia & Upjohn SA who also provided the recombinant human growth hormone (Genotonorm). We are grateful to all the parents and children who participated in the study. We also thank Dr D. Chevenne for measurements of serum insulin. Finally, we thank Mrs. Giraud and the nursing staff of the Centre d'Investigations Cliniques of the Hôpital Robert-Debré.

References

1 Sempé M, Pédron G, Pernot MP: Auxologie: Méthodes et séquences. Paris, Theraplix, 1979.

2 Greulich WW, Pyle SI (eds): Radiographic Atlas of Skeletal Development of the Hand and Wrist, ed 2. Stanford, Stanford University Press, 1959.

3 Haarbo J, Gotfredsen A, Hassager, Christiansen C: Validation of body composition measurement by dual-energy X-ray absorptiometry (DEXA). Clin Physiol 1991;11:331–341.

4 Root AW, Bongiovanni AM, Eberlin WR: Studies of the secretion and metabolic effects of human growth hormone in children with glucocorticoid-induced growth retardation. J Pediatr 1969;75:826–832.

5 Sadeghi-Nejad A, Senior B: Adrenal function, growth, and insulin in patients treated with corticoids on alternate days. Pediatrics 1969;43: 277–283.

6 Morris HG: Growth and skeletal maturation in asthmatic children: Effect of corticosteroid treatment. Pediatr Res 1975;9:579–583.

7 Tenore A, Berman WF, Parks JS, Bongiovanni AM. Basal and stimulated serum growth hormone concentrations in inflammatory bowel disease. J Clin Endocrinol Metab 1977;44: 622–628.

8 Kaufman S, Jones KL, Wehrenberg WB, Culler FL. Inhibition by prednisone of growth hormone (GH) response to GH-releasing hormone in normal men. J Clin Endocrinol Metab 1988; 67:1258–1261.

9 Pralong FP, Miell JP, Corder R, Gaillard RC: Dexamethasone treatment in man induces changes in 24 hour growth hormone (GH) secretion profile without altering total GH released. J Clin Endocrinol Metab 1991;73: 1191–1196.

10 Butenandt O: Rheumatoid arthritis and growth retardation in children: Treatment with human growth hormone. Eur J Pediatr 1979;130:15–28.

11 Sturge RA, Beardwell C, Hartog M, Wright D, Ansell BM: Cortisol and growth hormone secretion in relation to linear growth: Patients with Still disease on different therapeutic regimens. Br Med J 1970;iii:547–551.

12 Allen RC, Jimenez M, Cowell CT: Insulin-like growth factor and growth hormone secretion in juvenile chronic arthritis. Ann Rheum Dis 1991;50:602–606.

13 Davies UM, Rooney M, Preece MA, Ansell BM, Woo P: Treatment of growth retardation in juvenile chronic arthritis with recombinant human growth hormone. J Rheumatol 1994; 21:153–158.

14 Ward DJ, Hartog M, Ansell BM. Corticosteroid-induced dwarfism in Still's disease treated with human growth hormone: Clinical and metabolic effects including hydroxyproline excretion in two cases. Ann Rheum Dis 1966;25: 416–421.

15 Svantesson H: Treatment of growth failure with human growth hormone in patients with juvenile chronic arthritis: A pilot study. Clin Exper Rheumatol 1991;9(suppl 6):47–50.

16 Allen D, Goldberg B: Stimulation of collagen synthesis and linear growth by growth hormone in glucocorticoid-treated children. Pediatrics 1992;89:416–421.

17 Davies UM, Jones J, Reeve J, Camacho-Hubner C, Charlett A, Ansell BM, Preece MA, Woo PMM: Juvenile chronic arthritis: Effects of disease activity and recombinant human growth hormone on insulin-like growth factor-1, insulin-like growth factor-binding proteins 1 and 3, and osteocalcin. Arthritis Rheum 1997;40: 332–340.

18 Bentham J, Rodriguez-Arnao J, Ross RJM: Acquired growth hormone resistance in patients with hypercatabolism. Horm Res 1993;40:87–91.

19 Clemmons DR: Role of insulin-like growth factor-1 in reversing catabolism. J Clin Endocrinol Metab 1992;75:1183–1185.

20 Gourmelen M, Girard F, Binoux M: Serum somatomedin/insulin-like factor (IGF) and IGF carrier levels in patients with Cushing's syndrome or receiving glucocorticoid therapy. J Clin Endocrinol Metab 1982;54:885–892.

21 Elders MJ, Wingfield BS, McNatt ML, Clarke JS, Hughes ER: Glucocorticoid therapy in children: Effect on somatomedin secretion. Am J Dis Child 1975;129:1393–1396.

22 Léger J, Garel C, Legrand I, Paulsen A, Hassan M, Czernichow P: Magnetic resonance imaging evaluation of adipose tissue and muscle tissue mass in children with growth hormone deficiency, Turner's syndrome and intrauterine growth retardation during the first year of treatment with GH. J Clin Endocrinol Metab 1994; 78:904–909.

23 Jorgensen JOL, Vahl N, Hansen TB, Thuesen L, Hagen C, Christiansen JS: Growth hormone versus placebo treatment for one year in growth hormone deficient adults: Increase in exercise capacity and normalization of body composition. Clin Endocrinol (Oxf) 1996;45:681–688.

24 Touati G, Prieur AM, Ruiz JC, Noel M, Czernichow P: Beneficial effects of one-year growth hormone administration to children with juvenile chronic arthritis on chronic steroid therapy. I. Effects on growth velocity and body composition. J Clin Endocrinol Metab 1998;83: 403–409.

25 Vandeweghe M, Taelman P, Kaufman JM: Short- and long-term effects of growth hormone treatment on bone turnover and bone mineral content in adult growth hormone-deficient males. Clin Endocrinol (Oxf) 1993;39: 409–415.

26 Degerblad M, Elgindy N, Hall K, Sjöberg HE, Thoren M: Potent effect of recombinant growth hormone on bone mineral density and body composition in adults with panhypopituitarism. Acta Endocrinol (Copenh) 1992;126: 387–393.

27 O'Halloran DJ, Tsatsoulis A, Whitehouse RW, Holmes SJ, Adams JE, Shalet SM: Increased bone density after recombinant human growth hormone (GH) therapy in adults with isolated GH deficiency. J Clin Endocrinol Metab 1993; 76:1344–1348.

28 Bratusch-Marrain PR, Smith D, DeFronzo RA: The effect of growth hormone on glucose metabolism and insulin secretion in man. J Clin Endocrinol Metab 1982;55:973–982.

29 Botero D, Danon M, Brown RS: Symptomatic non-insulin dependent diabetes mellitus during therapy with recombinant human growth hormone. J Pediatr 1993;123:590–592.

30 Bowes SB, Umpleby M, Cummings MH, Jackson NC, Carroll PV, Lowy C, Sönksen PH, Russel-Jones DL: The effect of recombinant human growth hormone on glucose and leucine metabolism in Cushing's syndrome. J Clin Endocrinol Metab 1997;82:243–246.

Hormone Research

Horm Res 2000;53(suppl 1):87–97

Growth Hormone and Adipocyte Function in Obesity

Su Youn Nam[a] Claude Marcus[b]

[a]Division of Endocrinology, Department of Internal Medicine, Yonsei University College of Medicine, Seoul, Korea; [b]Department of Pediatrics, Endocrine Research Unit, Karolinska Institute, Huddinge University Hospital, Stockholm, Sweden

Key Words

Growth hormone · Insulin-like growth factor I · Growth hormone binding protein · Adipocyte · Obesity

Abstract

In obesity, growth hormone (GH) secretion is impaired which is considered a consequence rather than a cause of obesity. GH regulates the expression of GH receptor and the synthesis of insulin-like growth factor I (IGF-I) in adipocytes. Although GH hyposecretion in obesity may decrease the generation of IGF-I in each adipocyte, increased amounts of IGF-I and GH-binding protein could be secreted from the excessively enlarged amounts of adipose tissue. This may contribute to the normal/high serum-IGF-I and high GH-binding protein levels in obesity. Hyperinsulinemia and increased GH receptor activity may also affect the GH-IGF-I axis. Favorable effects of GH treatment have been observed in obese children and adults. GH treatment decreases adiposity, reduces triglyceride accumulation by inhibiting lipoprotein lipase and enhances lipolysis both via increased hormone-sensitive lipase activity and via induction of beta adrenoreceptors. GH treatment also has a favorable effect on obesity-associated dyslipidemia, but the effects on insulin sensitivity have been conflicting.

Introduction

Prevalence and severity of childhood obesity are increasing rapidly worldwide [1]. Approximately half of obese schoolchildren remain obese as adults [2], and, once established in adulthood, obesity can rarely be resolved through voluntary weight loss [3]. In particular, an excessive deposition of intra-abdominal fat has been known to be associated with an increase in various metabolic complications such as glucose intolerance, hyperlipidemia and cardiovascular disorders [4–6]. In addition, both spontaneous [7–9] and stimulated [10–13] growth hormone (GH) secretions are impaired in obesity which is reversible upon weight loss [14–16] and thus is considered a consequence rather than the cause of the obese state. However, despite the low GH levels, normal [17–20] or increased [21–24] insulin-like growth factor (IGF) I levels have been observed in obese subjects. Indeed, growth velocity is commonly normal or increased in obese children [25]. As for the possible mechanisms underlying these observations, overnutrition [26–29], chronic hyperinsulinemia [30] and high free IGF-I [18, 20, 24] have been suggested.

Considering that adipose tissue is a well-documented target tissue for GH [31–33], GH has been proposed to possess a role in the regulation of metabolic effects on lipolysis and lipogenesis [34], as well as in differentiation of the preadipocyte [35]. In accordance with these actions

KARGER

Fax + 41 61 306 12 34
E-Mail karger@karger.ch
www.karger.com

0301–0163/00/0537–0087$17.50/0

Accessible online at:
www.karger.com/journals/hre

Prof. C. Marcus
Pediatric Endocrine Research Unit, B62
Karolinska Institute, Huddinge University Hospital
S–141 86 Huddinge (Sweden)
Tel. +46 8 585 81419, Fax +46 8 585 87370, E-Mail claude.marcus@pediat.hs.sll.se

on adipocytes, GH treatment markely decreases adipose tissue mass. GH secretion is sharply reduced in obesity, with the consequent loss of the lipolytic effect that would be expected if the fat cells were exposed to normal levels of circulating GH. This has led to the suggestion of a possible therapeutic role for GH in obesity.

In this review, we discuss the deranged GH-IGF-I axis and its relationship with GH-binding proteins (GHBP)/GH receptor observed in obesity. In addition, we will summarize the actions of GH on adipocytes as well as the effect of GH treatment on obesity.

GH-IGF-I Axis and GHBP/GH Receptor in Obesity

Mechanisms of GH Secretory Dysfunction in Obesity

Obesity-related impairment of GH secretion, either spontaneous [7–9] or evoked by all provocative stimuli [10–13], has attracted considerable study. Although enhancement of GH clearance is a contributing factor, downregulation of GH secretion in obesity can be explained by three mechanisms: (1) a decrease in GH-releasing hormone (GHRH) release or an increase in somatostatin activity from the hypothalamus; (2) decreased responsiveness of the pituitary somatotrophs and (3) a perturbation of the peripheral signals at either the pituitary or the hypothalamic level.

A lower number of spontaneous GH secretory bursts observed in obesity [7, 8] allows for speculation that a GHRH secretory defect could be the cause of GH hyposecretion. In genetically obese Zucker rats, GHRH content and expression, as well as pituitary GH stores, were decreased [36–38] which suggests that a decrease in GH secretion in obesity, as a consequence of a decreased GHRH release from the hypothalamus, cannot be excluded.

On the basis of studies that showed that pharmacological agents thought to reduce somatostatinergic activity (pyridostigmine, arginine, galanin, beta-adrenergic antagonists) improve GHRH-stimulated GH secretion in obese subjects [11, 39–45], an enhanced somatostatinergic tone could be the cause of obesity-related GH secretory dysfunction. However, GH responses are lower than those observed in normal-weight subjects, even if the somatostatinergic activity is suppressed with these drugs [11, 41, 44, 45], which indicates that this higher somatostatinergic tone is unable to fully explain the GH hyposecretion in obesity.

Free fatty acids (FFA) participate in the regulation of GH secretion, acting at the pituitary [46] and the hypothalamus by enhancing somatostatin release [47]. Decreased plasma FFA levels stimulate GH secretion, and increased levels block GH secretion elicited by all stimuli (GHRH, pyridostigmine, GHRH plus pyridostigmine) [11–13]. In obese subjects, GH release is improved to levels similar to those of normal subjects by an acute reduction of circulating FFA levels induced by acipimox, a nicotinic acid analogue that is able to block lipolysis [11–13, 48–50]. However, when lipolysis is blocked to a similar extent by acipimox, the GH response is lower compared with that in normal subjects [48–50]. Considering that obesity is characterized by high circulating FFA levels [51], chronic elevation in FFA levels could be an important factor in regulating the magnitude of GH secretion in obesity. Indeed, the prolonged suppression of FFA levels by chronic treatment with acipimox further improves the GH responses to stimuli [13, 50]. Alternatively, the observation that GH secretion is increased during fasting both in obese and in normal weight subjects despite predicted inhibitory effects of elevated plasma FFA levels with fasting [52] has been contradictory to the possible role of FFA in regulating GH secretion in obesity.

All of the observed abnormalities are reversed with a significant reduction in body weight which suggests that an impairment of GH secretion in obesity seems to be a metabolic consequence of the obese state.

Effect of Obesity on GH-IGF-I Axis and GHBP

Despite GH hyposecretion in obesity and the subsequent potential abnormalities in the peripheral GH-IGF-I axis, prepubertal obese children are generally of normal-to-tall stature [25]. Indeed, total serum IGF-I levels in obesity have been reported to be normal or high, and increased circulating free IGF-I levels have consistently been observed in obese subjects. A similar discordance between GH secretion and IGF-I levels is observed in malnutrition due to anorexia nervosa [29]. In these patients, total IGF-I levels are markedly decreased regardless of GH secretory profiles [26–29]. Taken together, these observations suggest that, in overnutrition and extreme undernutrition, the IGF-I system seems to be largely controlled by factors other than GH, including nutritional factors. Although chronic overnutrition in obesity could stimulate IGF-I production [26], short-term overfeeding fails to normalize IGF-I in GH-deficient patients which suggests that an excess of nutrient intake is not likely to override the effects of GH deficiency to normalize serum IGF-I levels [27]. Furthermore, overnutrition is not as potent a stimulus for increasing IGF-I as nutritional restriction is for decreasing it, at least in adults, as obese subjects maintain their IGF-I level during diet restriction supplied with sufficient protein [28, 53, 54]. In malnourished patients, an observed low IGF-I level despite increased GH secretion indicates GH resistance may exist [29, 55].

Insulin modulates the rat liver GH receptor and stimulates the generation of IGF-I RNA in primary culture of hepatocytes, even in the absence of GH [56, 57], which suggests chronic hyperinsulinemia observed in obesity could play a role in maintaining normal IGF-I levels. In the main, IGF-I in the circulation is bound to IGF-binding protein (IGFBP) 3, whereas the other circulating binding proteins (IGFBP-1, IGFBP-2 and IGFBP-4) are believed to bind the remainder of IGF-I in serum [58–60]. Lower levels of IGFBP-1 and IGFBP-2 observed in obesity have been reported to have a negative correlation with the serum insulin level [61–63], suggesting that insulin is an important regulator of the production of IGFBP-1. On the other hand, the demonstration that no correlations were found between basal insulin level and IGFBP-1 in nonobese adults [64], as well as in fasting normoinsulinemic obese children [65], suggested that decreased IGFBP-1 levels in obesity cannot be fully explained by hyperinsulinemia. Moreover, under the controlled hypoinsulinemic conditions, the response of IGFBP-1 to insulin withdrawal is blunted, and the IGFBP-1 level remains low in obese subjects as compared with normal-weight subjects [62]. In patients with polycystic ovarian syndrome characterized by insulin resistance, insulin sensitivity mainly contributes to the differences in IGFBP-1 levels [63]. These findings indicate that low IGFBP-1 levels may be a compensatory response to decreased insulin sensitivity rather than a consequence of hyperinsulinemia, as reduced IGFBP-1 levels allow an increase in tissue availability of IGF-I which in turn stimulates glucose uptake in obese subjects with insulin resistance.

A significant decrease in both IGFBP-1 and IGFBP-2, as well as an increase in the total IGF-I level, could increase the portion of IGF-I in the free, biologically active form. More biologically active IGF-I could feedback to reduce GH release. Indeed, the absolute level of free IGF measured by the assays currently available is inaccurate, because no method has satisfactorily eliminated the problem of the loss of equilibrium that occurs during the analysis procedures. Therefore, the IGF/IGFBP molar ratio appears to be a more accurate indicator of circulating IGF delivered from their binding proteins [19]. The higher IGF/IGFBP molar ratio observed in obese children could explain the sustained growth.

Increased circulating GHBP levels have been observed in obese subjects with positive correlation between GHBP levels and body mass index and amount of body fat [66–69]. Considering that the GHBP is mainly generated from proteolysis of the extracellular domain of the GH receptor, changes in circulating GHBP levels may reflect changes in GH receptor levels in target tissues [70, 71]. These observations suggest that increased GHBP levels reflect GH hypersensitivity in order to compensate for GH hyposecretion in obesity. This, in turn, could partially explain that neither IGF-I nor IGFBP-3 levels are decreased in obesity despite GH hyposecretion.

The demonstration that IGF-I mRNA levels in adipose tissue are similar to those found in the liver [72], known to be the major source of circulating IGF-I, has led to the hypothesis that adipose tissue could be a major contributor of circulating IGF-I levels in obesity (fig. 1). In adipocytes cultured with GH, both IGF-I mRNA and, as a consequence, IGF-I levels in the culture medium are increased [73, 74]. GH is apparently the main regulator of IGF-I mRNA expression in adipose tissue. Hypophysectomy results in a marked decrease in IGF-I mRNA in rat adipose tissues. GH replacement almost completely restored IGF-I mRNA level and normalized the IGF-I level in this tissue [72, 73]. Although GH hyposecretion in obesity may cause a decrease in the generation of IGF-I in each adipocyte, significantly greater amounts of IGF-I could be secreted from the accumulated adipose tissues in obesity than from other tissues on a tissue weight basis [75]. This may contribute to the normal IGF-I level in the circulation in obesity. From this standpoint, the low GH secretion in obesity is caused by a physiological negative feedback by high IGF-I levels and not by GH dysregulation.

Moreover, recently, short isoforms of the human GH receptor, physiologically produced by alternative splicing, have been demonstrated [76–79] (fig. 2). Compared with the expression of the human exon-3-deleted GH receptor, which is regulated individually [80], the expression of these truncated isoforms seems to be tissue specific [76, 77]. Differential tissue expression of the truncated receptor with low abundance of the alternative splice in liver, kidney and fibroblasts, but similar expression to full-length receptor in mammary gland and adipose tissue, has been reported [76, 77]. This differential tissue expression of the alternatively spliced transcript may regulate the production of GHBP (fig. 1). Functional studies have shown that truncated GH receptor exhibits impaired internalization [76, 77] and, therefore, will allow more receptors to be available for proteolysis. The expression of

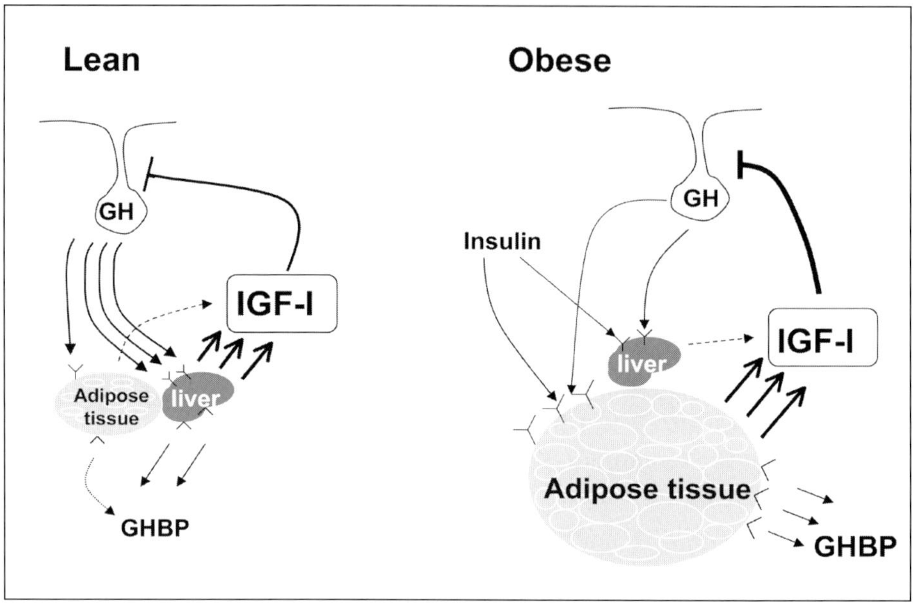

Fig. 1. Although GH hyposecretion in obesity may decrease IGF-I production in the liver and in each adipocyte, excessive accumulation of adipose tissue in obesity can lead to greater amounts of secreted IGF-I. In addition, increased insulin levels observed in obesity stimulate the expression of GH receptor as well as IGF-I production. Increased expression of alternative splicing in adipose tissue may contribute to an elevated circulating GHBP level.

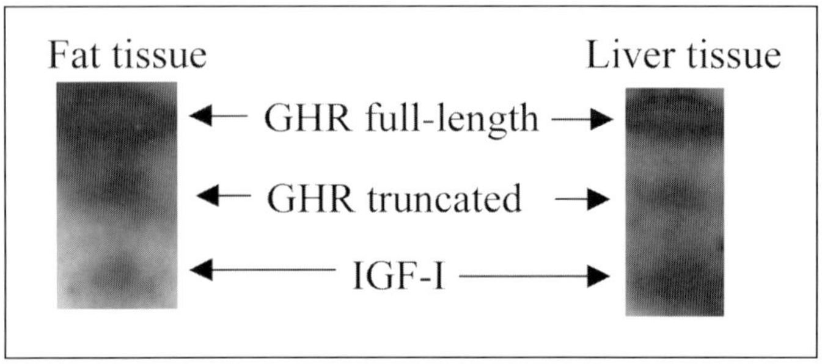

Fig. 2. Expression of full-length GH receptor (GHR), truncated GHR and IGF-I mRNA by RNAse protection analysis of human adipose tissue and liver.

truncated GH receptor in adipose tissue may contribute to an elevated circulating GHBP level in obesity (fig. 1).

Effect of GH on Adipocytes

Biological Effects of GH on Adipocytes

Expression of the GH receptor in adipose tissue has long been demonstrated [31–33], resulting in the identification of adipose tissue as a direct target tissue for GH. It is apparent that there is differential expression of the receptor, depending on the location of the adipose tissue depot [81]. For example, in the rat, a high level of expression of the GH receptor is observed in the epididymal fat pad, whereas the retroperitoneal fat pad in comparison expresses a low level of GH receptor [81]. Such location-dependent expression of the GH receptor may explain the ability of GH administered to adult obese subjects to selectively reduce the abdominal fat mass in preference to other fat depots [82–84]. In the human, both the exon-3-retaining and exon-3-deleting isoforms of the GH receptor are expressed [80]. Hypophysectomy has been demonstrated to decrease adipose tissue GH receptor gene expression in the rat, and GH replacement restores receptor expression towards normal [72, 85]. Similarly in the human, treatment of Prader-Willi syndrome or GH-deficient patients with GH results in an upregulation of adipose tissue GH receptor gene expression [86]. Thus, homologous upregulation of the adipose tissue GH receptor is observed at least under absolute or relative hyposomatotropic conditions.

GH treatment of patients with GH deficiency as well as idiopathic short-stature children increases circulating GHBP levels [87, 88]. These findings suggest that GH treatment increases the sensitivity to GH, as the dose of GH can be reduced after a couple of months of treatment, without any negative effects on IGF-I levels or on other metabolic parameters in GH-deficient adults [82].

On the basis of extensive studies using preadipocyte cell lines, such as the 3T3-F442A [89, 90], 3T3-L1 [91], Ob1771 and Ob17UT [92], it has been demonstrated that GH promotes their differentiation. In contrast, GH exhibits a potent inhibitory activity on adipocyte differentiation in primary preadipocyte culture directly obtained from the rat [93, 94] or human [35]. The contradictory findings are probably a result of differences in the stage of commitment and differentiation of the cells used in the different studies.

Recently, several factors have been identified that regulate preadipocyte differentiation at the molecular level. Adipocyte determination and differentiation factor 1 [95, 96], peroxisome proliferator activated receptor gamma [97–99], CCAAT/enhancer-binding proteins alpha, beta and delta [100], adipocyte P2 [101] and FFA synthetase [102, 103] have all been found to be involved at different stages. GH inhibits the differentiation of primary rat preadipoctye at a step after the induction of adipocyte determination and differentiation factor 1, but before the induction of adipocyte P2 and FFA synthetase [104]. The mechanism by which GH inhibits the differentiation [104] is not dependent upon MAP kinase activated serine phosphorylation of peroxisome proliferator activated receptor gamma, as has been demonstrated in studies with other growth factors [105, 106]. The precise molecular mechanisms involved in inhibitory action of GH on differentiation of human adipocyte are still unknown and require further investigation.

On the other hand, GH stimulates the production of IGF-I which, in turn, stimulates preadipocyte proliferation in an autocrine/paracrine manner, leading to an enlarged pool of adipocytes [107, 108]. Children with GH deficiency have an increased mean adipocyte volume, but a reduced number of fat cells as compared with normal children, and GH replacement normalizes these changes [109]. Although the net effect of systemic GH treatment has not been defined regarding contradictory actions of GH and IGF-I on adipocyte differentiation, GH treatment of obese subjects significantly reduces adiposity. These clinical findings are not only the result of the biological effects of GH and IGF-I in adipose tissue but also of their effects on other target tissues, such as CNS and muscle cells.

Metabolic Effects of GH on Adipocytes

GH exerts a variety of metabolic effects either directly or indirectly on adipocytes. GH has opposing effects on glucose and lipid metabolism in adipose tissue: insulin-like effects and insulin-antagonizing effects, depending on the time frame [110]. The insulin-like effect is an acute antilipolytic and lipogenic effect, whereas the long-term insulin-antagonizing effect inhibits lipogenesis and glucose transport and increases lipolysis [111].

Acute and transient insulin-like actions of GH on the adipose tissue are observed in vitro in rat adipocytes [112, 113] but not in human adipocytes [114]. Adipose tissue from the rat preincubated without GH for several hours shows an increased glucose oxidation and antilipolysis when GH is subsequently added [115]. The metabolic significance is not clear, but phosphorylation and activation of some insulin-sensitive enzymes, IRS-I and IRS-2, and activation of PI3 kinase may result in activation of a common signaling pathway to promote glucose uptake and inhibition of lipolysis [116].

GH may modify glucose disposal directly or indirectly via induction of IGF-I. Of note, the IGF-I receptor is not expressed in the mature adipocyte [117], and, thus, the effect of systemic GH treatment on adipocyte function is not mediated by IGF-I. GH treatment in vivo [118, 119] or longer exposure of GH in vitro [120–124] induces a suppression of glucose metabolism in adipose tissues in the presence or absence of insulin. GH acts at multiple sites on insulin-signaling pathways to induce insulin resistance. It has been demonstrated that GH inhibits basal and insulin-stimulated phosphorylation of the insulin receptor [125] as well as that GH reduces the number of GLUT1 transporters and GLUT1 mRNA, with no change or a decrease in GLUT4 in adipocyte plasma membrane [124, 125].

The average size and lipid content of subcutaneous adipocytes in GH-deficient patients is greater than in normal subjects [109, 126], while GH treatment results in decreased size and lipid content of their adipocytes. The lipolytic effect of GH (fig. 3) is well documented both in vivo [127, 128] and in vitro [129, 130]. In hypophysectomized rats, GH treatment restores the decrease in catecholamine-induced lipolysis and also the number of beta adrenoreceptors [129, 130]. In adults with GH deficiency, GH treatment increases the responsiveness of epinephrine-induced lipolysis [127]. In adipocytes isolated from obese children after chronic GH treatment, maximal catecholamine-induced lipolysis is increased [128]. This increased lipolytic sensitivity of adipocytes is induced by terbutaline and not by isoprenaline, which indicates that the lipolytic effect of GH is related to a stimulatory effect on the β_2-adrenoreceptor in adipocytes (fig. 3) [128].

GH inhibits phosphodiesterase, which converts cAMP to AMP, thereby activating the cAMP-dependent protein kinase and hormone-sensitive lipase, resulting in a decrease in fat storage by hydrolysis of triglycerides to glycerol and FFAs [130].

In contrast, data concerning the effect of systemic GH treatment on lipogenesis in adipocytes have been conflicting. In human adipocytes isolated after systemic GH treatment, basal lipogenesis increases [127, 128, 131],

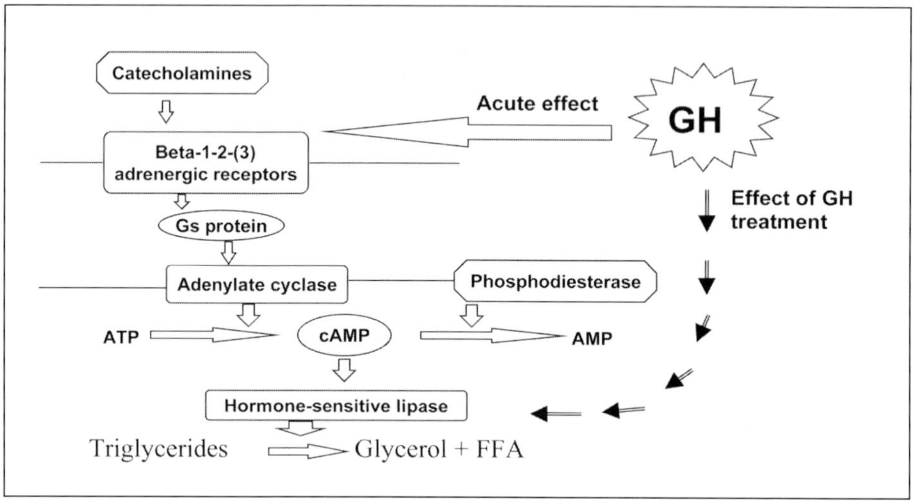

Fig. 3. Effect of GH on the lipolytic cascade in human adipocytes. GH increases lipolysis, with hormone-sensitive lipase activity being increased and via induction of β-adrenoreceptor and phosphodiesterase.

whereas insulin-induced lipogenesis is increased in Prader-Willi syndrome [131] and GH-deficient children [Marcus, unpubl. data, 1999], but it is not changed in obese children [128].

The accumulation of triglycerides in human adipose tissue mainly depends on lipoprotein lipase (LPL). LPL induces the release of FFA from triglycerides transported as circulating very-low-density lipoprotein triglycerides and chylomicrons. The increased uptake of FFAs into adipose tissues stimulates the accumulation of lipid in adipocytes [132]. In both adipocytes incubated with GH [111] and in human adipocytes with systemic GH treatment [133, 134], GH reduces the activity of LPL. The GH-induced suppression in LPL activity differs in various depots which could, in part, be responsible for a more pronounced reduction in intra-abdominal fat depots than in subcutaneous fat after GH treatment [135, 136].

Effect of GH Treatment on Obesity

GH-deficient patients exhibit an increase in the amount of body fat, especially marked deposition of intra-abdominal fat in adults, which is known to be associated with increased cardiovascular morbidity and mortality [137, 138]. In these GH-deficient patients, GH replacement has demonstrated favorable effects on most metabolic disturbances [82, 138–141]. Decreased GH secretion in obesity results in the consequent loss of a lipolytic effect of GH. Furthermore, insulin resistance and hyperinsulinemia observed in obesity cause insulin-induced antilipolytic actions.

Dietary restriction as a basic treatment for obesity is complicated by protein catabolism which results in a negative nitrogen balance and a loss of lean body mass [142]. GH exerts an anabolic action directly and indirectly by inducing IGF-I. Based on both lipolytic and anabolic actions of GH, a possible therapeutic role for GH in obesity can be expected. However, an effect of GH treatment on body composition in obesity has not consistently been observed [82, 83, 143–148]. A critical evaluation of the effects of GH on body composition has to take several points into account.

The timing and method used to assess the effect of GH should be considered. GH causes salt and water retention by a renal action that appears to depend on activation of the renin-angiotensin system [149]. It occurs within 4–6 weeks after the beginning of GH treatment, and thereafter any further changes in total body water are negligible. Therefore, the prolonged duration of GH treatment of at least 6 weeks and proper methods to evaluate the changes in total body fluid status are required in order to avoid misleading information on body composition.

The dose of GH can alter the effectiveness of GH treatment. Higher doses of GH cause more excretion of urinary-C-peptide which has a negative correlation with body fat loss [147]. This finding suggests that high doses of GH treatment induce a significant hyperinsulinemia which may antagonize the lipolytic effect of GH in obese adults.

Anabolic effects of GH have been demonstrated in studies of GH treatment combined with diet restriction [84, 134, 145, 148]. The nitrogen-sparing effects of GH depend on both the dose of GH and the degree of diet restriction.

The demonstration that GH treatment can improve the metabolic perturbations related to intra-abdominal adiposity [83] has attracted great interest. Abdominal obesity is associated with high levels of serum low-density lipoprotein and very low density liporptoein cholesterols as well as low levels of high-density lipoprotein cholesterol which could be an important risk factor for cardiovascular morbidity and mortality [4–6]. GH-deficient patients exhibit similar blood lipoprotein profiles [138–141] which normalize in response to GH replacement [82, 141]. In agreement with this finding, GH treatment in obesity results in a decrease in both low-density lipoprotein cholesterol and triglyceride levels. GH directly induces the hepatic apo-B/E receptor, leading to enhanced removal of both low-density and very low density lipoproteins from blood [150]. This direct effect of GH on lipid profiles, independent of any change in adiposity, can contribute to a reduced risk of cardiovascular disease associated with obesity.

Abdominal adiposity has been demonstrated to cause insulin resistance which is thought to be due to increased circulating FFA levels. Increased circulating FFA levels decrease hepatic insulin clearance and inhibit insulin-mediated glucose uptake by muscle which results in insulin resistance and hyperinsulinemia. This, in turn, leads to the tempting speculation that a significant reduction in visceral fat with GH treatment improves insulin sensitivity. However, results have been conflicting regarding the effect of GH on insulin sensitivity [83, 84, 129, 144]. Neither serum FFA nor insulin levels fall in response to a GH-induced decrease in visceral adiposity, even though an improvement in insulin-mediated glucose disposal has been reported [83; Nam, unpubl. data, 1999]. The mechanisms underlying this discrepancy are not yet clarified. The observations in obese adults that a higher dose and a shorter duration of GH treatment cause a marked insulin resistance and a significant increase in serum insulin levels [134, 144, 146] suggest that the dose and the duration of GH treatment can affect insulin resistance. In contrast, when combined with diet restriction, the GH dose does not induce a significant increase in insulin levels [84, 143]. On the other hand, GH treatment increases resting energy expenditure in GH-deficient patients [151] and prevents diet-induced reduction of resting energy expenditure in obese subjects [144]. Taken together, GH treatment, when combined with diet restriction, could be a more effective therapy in obesity.

Conclusions

In obesity, both spontaneous and stimulated GH secretions are impaired which are reversible upon weight loss and, thus, are considered a consequence rather than the cause of the obese state. Despite GH deficiency, normal or increased IGF-I levels have been observed. Adipose tissue is a direct target tissue for GH and could be a major source of circulating IGF-I and GHBP. GH treatment markely decreases adipose mass by increasing lipolysis and decreasing LPL activity. Decreased GH secretion in obesity results in the consequent loss of a lipolytic effect of GH which suggests a possible therapeutic role of GH in obesity. GH has a direct favorable effect on dyslipidemia associated with obesity. Further studies should be conducted to define optimal dose and duration of GH treatment that have favorable effects on body composition as well as on metabolic disturbances associated with obesity.

References

1 WHO: Prevention and management of the global epidemic of obesity. Report of the WHO consultation on obesity. Geneva, WHO, 1997.

2 Serdula MK, Ivery D, Coates RJ, Freedman DS, Williamson DF, Byers T: Do obese children become obese adults? A review of the literature. Prev Med 1993;22:167–177.

3 Stunkard A, Coll M, Lundquist S, Meyers A: Obesity and eating style. Arch Gen Psychiatry 1980;7:1127–1129.

4 Marin P, Björntorp P: Endocrine-metabolic pattern and adipose tissue distribution. Horm Res 1993;39(suppl 3):81–85.

5 Reaven GM: Pathophysiology of insulin resistance in human disease. Physiol Rev 1995;75: 473–486.

6 Larsson B, Svardsudd K, Welin L, Wilhelmsen L, Björntorp P, Tibblin G: Abdominal adipose tissue distribution, obesity, and risk of cardiovascular disease and death: 13 year follow up of participants in the study of men born in 1913. BMJ 1984;288:1401–1404.

7 Iranmanesh A, Lizarralde G, Veldhuis JD: Age and relative adiposity are specific negative determinants of the frequency and amplitude of growth hormone (GH) secretory bursts and the half-life of endogenous GH in healthy men. J Clin Endocrinol Metab 1991;73:1081–1088.

8 Veldhuis JD, Liem AY, South S, Weltman A, Weltman J, Clemmons DA, Abbott R, Mulligan T, Johnson ML, Pincus S, Straume M, Iranmanesh A: Differential impact of age, sex steroid hormones, and obesity on basal versus pulsatile growth hormone secretion in men as assessed in an ultrasensitive chemiluminescence assay. J Clin Endocrinol Metab 1995;80: 3209–3222.

9 Weltman A, Weltman JY, Hartman ML, Abbott RD, Rogol AD, Evans WS, Veldhuis JD: Relationship between age, percentage body fat, fitness, and 24-hour growth hormone release in healthy young adults: Effects of gender. J Clin Endocrinol Metab 1994;78:543–548.

10 Andreotti AC, Lanzi R, Manzoni MF, Caumo A, Moreschi A, Pontiroli AE: Acute pharmacologic blockade of lipolysis normalizes nocturnal growth hormone levels and pulsatility in obese subjects. Metabolism 1994;43:1207–1213.
11 Lee EJ, Nam SY, Kim KR, Lee HC, Cho JH, Nam MS, Song YD, Lim SK, Huh KB: Acipimox potentiates growth hormone (GH) response to GH-releasing hormone with or without pyridostigmine by lowering serum free fatty acid in normal and obese subjects. J Clin Endocrinol Metab 1995;80:2495–2498.
12 Nam SY, Lee EJ, Kim KR, Lee HC, Nam MS, Cho JH, Huh KB: Long-term administration of acipimox potentiates growth hormone response to growth hormone-releasing hormone by decreasing serum free fatty acid in obesity. Metabolism 1996;45:594–597.
13 Pontiroli AE, Manzoni MF, Malighetti ME, Lanzi R: Restoration of growth hormone (GH) response to GH-releasing hormone in elderly and obese subjects by acute pharmacological reduction of plasma free fatty acids. J Clin Endocrinol Metab 1996;81:3998–4001.
14 Williams T, Berelowitz M, Joffe SN, Thorner MO, Rivier J, Vale W, Frohman LA: Impaired growth hormone responses to growth hormone-releasing factor in obesity. A pituitary defect reversed with weight reduction. N Engl J Med 1984;311:1403–1407.
15 Tanaka K, Inoue S, Numata K, Okazaki H, Nakamura S, Takamura Y: Very-low-calorie diet-induced weight reduction reverses impaired growth hormone secretion response to growth hormone-releasing hormone, arginine, and *L*-dopa in obesity. Metabolism 1990;39: 892–896.
16 Rasmussen MH, Hvidberg A, Juul A, Main KM, Gotfredsen A, Skakkebaek NE, Hilsted J, Skakkebaek NE: Massive weight loss restores 24-hour growth hormone release profiles and serum insulin-like growth factor-I levels in obese subjects. J Clin Endocrinol Metab 1995; 80:1407–1415.
17 Slowinska-Srzednicka J, Zgliczynski W, Makowska A, Jeske W, Brzezinska A, Soszynski P, Zgliczynski S: An abnormality of the growth hormone/insulin-like growth factor-I axis in women with polycystic ovary syndrome due to coexistent obesity. J Clin Endocrinol Metab 1992;74:1432–1435.
18 Frystyk J, Vestbo E, Skjaerbaek C, Mogensen CE, Orskov H: Free insulin-like growth factors in human obesity. Metabolism 1995;44(suppl 4):37–44.
19 Radetti G, Bozzola M, Pasquino B, Paganini C, Aglialoro A, Livieri C, Barreca A: Growth hormone bioactivity, insulin-like growth factors (IGFs), and IGF binding proteins in obese children. Metabolism 1998;47:1490–1493.
20 Argente J, Caballo N, Barrios V, Pozo J, Munoz MT, Chowen JA, Hernandez M: Multiple endocrine abnormalities of the growth hormone and insulin-like growth factor axis in prepubertal children with exogenous obesity: Effect of short- and long-term weight reduction. J Clin Endocrinol Metab 1997;82:2076–2083.
21 Loche S, Cappa M, Borrelli P, Faedda A, Crino A, Cella SG, Corda R, Muller EE, Pintor C: Reduced growth hormone response to growth hormone-releasing hormone in children with simple obesity: Evidence for somatomedin-C mediated inhibition. Clin Endocrinol (Oxf) 1987;27:145–153.
22 Van Vliet G, Bosson D, Rummens E, Robyn C, Wolter R: Evidence against growth hormone-releasing factor deficiency in children with idiopathic obesity. Acta Endocrinol Suppl (Copenh) 1986;279:403–410.
23 Caufriez A, Golstein J, Lebrun P, Herchuelz A, Furlanetto R, Copinschi G: Relations between immunoreactive somatomedin C, insulin and T_3 patterns during fasting in obese subjects. Clin Endocrinol (Oxf) 1984;20:65–70.
24 Nam SY, Lee EJ, Kim KR, Cha BS, Song YD, Lim SK, Lee HC, Huh KB: Effect of obesity on total and free insulin-like growth factor (IGF)-1, and their relationship to IGF-binding protein (BP)-1, IGFBP-2, IGFBP-3, insulin, and growth hormone. Int J Obes Relat Metab Disord 1997;21:355–359.
25 Dietz WH Jr, Hartung R: Changes in height velocity of obese preadolescents during weight reduction. Am J Dis Child 1985;139:705–707.
26 Thissen JP, Ketelslegers JM, Underwood LE: Nutritional regulation of the insulin-like growth factors. Endocr Rev 1994;15:80–101.
27 Underwood LE, Smith EP, Clemmons DR, Maes M, Maiter D, Ketelslegers JM: The production and actions of insulin-like growth factors: Their relationship to nutrition and growth; in Tanner JM (ed): Auxology. London, Smith-Gordon, 1989, pp 235–249.
28 Smith SR, Edgar PJ, Pozefsky T, Chhetri MK, Prout TE. Growth hormone in adults with protein-calorie malnutrition. J Clin Endocrinol Metab 1974;39:53–62.
29 Hochberg Z, Hertz P, Colin V, Ish-Shalom S, Yeshurun D, Youdim MB, Amit T: The distal axis of growth hormone (GH) in nutritional disorders: GH-binding protein, insulin-like growth factor-I (IGF-I), and IGF-I receptors in obesity and anorexia nervosa. Metabolism 1992;41:106–112.
30 Zannolli R, Rebeggiani A, Chiarelli F, Morgese G: Hyperinsulinism as a marker in obese children. Am J Dis Child 1993;147:837–841.
31 Gavin JR 3rd, Saltman RJ, Tollefsen SE: Growth hormone receptors in isolated rat adipocytes. Endocrinology 1982;10:637–643.
32 Leung DW, Spencer SA, Cachianes G, Hammonds RG, Collins C, Henzel WJ, Barnard R, Waters MJ, Wood WI: Growth hormone receptor and serum binding protein: Purification, cloning and expression. Nature 1987;330:537–543.
33 Vikman K, Carlsson B, Billig H, Eden S: Expression and regulation of growth hormone (GH) receptor messenger ribonucleic acid (mRNA) in rat adipose tissue, adipocytes, and adipocyte precursor cells: GH regulation of GH receptor mRNA. Endocrinology 1991;129: 1155–1161.
34 Richelsen B: Action of growth hormone in adipose tissue. Horm Res 1997;48(suppl 5):105–110.
35 Wabitsch M, Hauner H, Heinze E, Teller WM: The role of growth hormone/insulin-like growth factors in adipocyte differentiation. Metabolism 1995;44(suppl 4):45–49.
36 Tannenbaum GS, Lapointe M, Gurd W, Finkelstein JA: Mechanisms of impaired growth hormone secretion in genetically obese Zucker rats: Roles of growth hormone-releasing factor and somatostatin. Endocrinology 1990;127: 3087–3095.
37 Ahmad I, Steggles AW, Finkelstein JA: In situ hybridization study of obesity-associated alteration in growth hormone mRNA levels. Int J Obes Relat Metab Disord 1992;16:435–441.
38 Ahmad I, Finkelstein JA, Downs TR, Frohman LA: Obesity-associated decrease in growth hormone-releasing hormone gene expression: A mechanism for reduced growth hormone mRNA levels in genetically obese Zucker rats. Neuroendocrinology 1993;58:332–337.
39 Massara F, Ghigo E, Molinatti P, Mazza E, Locatelli V, Muller EE, Camanni F: Potentiation of cholinergic tone by pyridostigmine bromide re-instates and potentiates the growth hormone responsiveness to intermittent administration of growth hormone-releasing factor in man. Acta Endocrinol (Copenh) 1986; 113:12–16.
40 Ghigo E, Mazza E, Imperiale E, Rizzi G, Benso L, Muller EE, Camanni F, Massara F: Enhancement of cholinergic tone by pyridostigmine promotes both basal and growth hormone (GH)-releasing hormone-induced GH secretion in children of short stature. J Clin Endocrinol Metab 1987;65:452–456.
41 Cordido F, Dieguez C, Casanueva FF: Effect of central cholinergic neurotransmission enhancement by pyridostigmine on the growth hormone secretion elicited by clonidine, arginine, or hypoglycemia in normal and obese subjects. J Clin Endocrinol Metab 1990;70: 1361–1370.
42 Loche S, Pintus S, Cella SG, Boghen M, Vannelli S, Benso L, Muller EE, Corda R, Pintor C: The effect of galanin on baseline and GHRH-induced growth hormone secretion in obese children. Clin Endocrinol (Oxf) 1990;33:187–192.
43 Loche S, Pintus S, Carta D, Muntoni AC, Congiu G, Civolani P, Pintor C: The effect of atenolol on the growth hormone response to growth hormone-releasing hormone in obese children. Acta Endocrinol (Oxf) 1992;126: 124–127.
44 Ghigo E, Procopio M, Boffano GM, Arvat E, Valente F, Maccario M, Mazza E, Camanni F: Arginine potentiates but does not restore the blunted growth hormone response to growth hormone-releasing hormone in obesity. Metabolism 1992;41:560–563.
45 Procopio M, Invitti C, Maccario M, Grottoli S, Cavagnini F, Camanni F, Ghigo E: Effect of arginine and pyridostigmine on the GHRH-induced GH rise in obesity and Cushing's syndrome. Int J Obes Relat Metab Disord 1995; 19:108–112.

46 Casanueva FF, Villanueva L, Dieguez C, Diaz Y, Cabranes JA, Szoke B, Scanlon MF, Schally AV, Fernandez-Cruz A: Free fatty acids block growth hormone (GH) releasing hormone-stimulated GH secretion in man directly at the pituitary. J Clin Endocrinol Metab 1987;65: 634–642.
47 Imaki T, Shibasaki T, Masuda A, Hotta M, Yamauchi N, Demura H, Shizume K, Wakabayashi I, Ling N: The effect of glucose and free fatty acids on growth hormone (GH)-releasing factor-mediated GH secretion in rats. Endocrinology 1986;118:2390–2394.
48 Cordido F, Fernandez T, Martinez T, Penalva A, Peino R, Casanueva FF, Dieguez C: Effect of acute pharmacological reduction of plasma free fatty acids on growth hormone (GH) releasing hormone-induced GH secretion in obese adults with and without hypopituitarism. J Clin Endocrinol Metab 1998;83:4350–4354.
49 Cordido F, Peino R, Penalva A, Alvarez CV, Casanueva FF, Dieguez C: Impaired growth hormone secretion in obese subjects is partially reversed by acipimox-mediated plasma free fatty acid depression. J Clin Endocrinol Metab 1996;81:914–918.
50 Maccario M, Procopio M, Grottoli S, Oleandri SE, Boffano GM, Taliano M, Camanni F, Ghigo E: Effects of acipimox, an antilipolytic drug, on the growth hormone (GH) response to GH-releasing hormone alone or combined with arginine in obesity. Metabolism 1996;45:342–346.
51 Ferrannini E: Physiological and metabolic consequences of obesity. Metabolism 1995; 44(suppl 3):15–17.
52 Kelijman M, Frohman LA: Enhanced growth hormone (GH) responsiveness to GH-releasing hormone after dietary manipulation in obese and nonobese subjects. J Clin Endocrinol Metab 1988;66:489–494.
53 Forbes GB, Brown MR, Welle SL, Underwood LE: Hormonal response to overfeeding. Am J Clin Nutr 1989;49:608–611.
54 Gama R, Teale JD, Marks V: The effect of synthetic very low calorie diets on the GH-IGF-1 axis in obese subjects. Clin Chim Acta 1990; 188:31–38.
55 Soliman AT, Hassan AE, Aref MK, Hintz RL, Rosenfeld RG, Rogol AD: Serum insulin-like growth factors I and II concentrations and growth hormone and insulin responses to arginine infusion in children with protein-energy malnutrition before and after nutritional rehabilitation. Pediatr Res 1986;20:1122–1130.
56 Johnson TR, Blossey BK, Denko CW, Ilan J: Expression of insulin-like growth factor I in cultured rat hepatocytes: Effects of insulin and growth hormone. Mol Endocrinol 1989;3:580–587.
57 Boni-Schnetzler M, Schmid C, Meier PJ, Froesch ER: Insulin regulates insulin-like growth factor I mRNA in rat hepatocytes. Am J Physiol 1991;260:E846–E851.
58 Brismar K, Hall K: Clinical applications of IGFBP-1 and its regulation. Growth Regul 1993;3:98–100.
59 Clemmons DR: Insulin-like growth factor binding proteins and their role in controlling IGF actions. Cytokine Growth Factor Rev 1997;8:45–62.
60 Kelley KM, Oh Y, Gargosky SE, Gucev Z, Matsumoto T, Hwa V, Ng L, Simpson DM, Rosenfeld RG: Insulin-like growth factor-binding proteins (IGFBPs) and their regulatory dynamics. Int J Biochem Cell Biol 1996;28:619–637.
61 Suikkari AM, Koivisto VA, Rutanen EM, Yki-Järvinen H, Karonen SL, Seppälä M: Insulin regulates the serum levels of low molecular weight insulin-like growth factor-binding protein. J Clin Endocrinol Metab 1988;66:266–272.
62 Conover CA, Lee PD, Kanaley JA, Clarkson JT, Jensen MD: Insulin regulation of insulin-like growth factor binding protein-1 in obese and nonobese humans. J Clin Endocrinol Metab 1992;74:1355–1360.
63 Morris DV, Falcone T: The relationship between insulin sensitivity and insulin-like growth factor-binding protein-1. Gynecol Endocrinol 1996;10:407–412.
64 Brismar K, Grill V, Efendic S, Hall K: The insulin-like growth factor binding protein-1 in low and high insulin responders before and during dexamethasone treatment. Metabolism 1991;40:728–732.
65 Saitoh H, Kamoda T, Nakahara S, Hirano T, Nakamura N: Serum concentrations of insulin, insulin-like growth factor (IGF)-I, IGF binding protein (IGFBP)-1 and -3 and growth hormone binding protein in obese children: Fasting IGFBP-1 is suppressed in normoinsulinaemic obese children. Clin Endocrinol (Oxf) 1998;48: 487–492.
66 Postel-Vinay MC, Saab C, Gourmelen M: Nutritional status and growth hormone-binding protein. Horm Res 1995;44:177–181.
67 Rasmussen MH, Ho KK, Kjems L, Hilsted J: Serum growth hormone-binding protein in obesity: Effect of a short-term, very low calorie diet and diet-induced weight loss. J Clin Endocrinol Metab 1996;81:1519–1524.
68 Roelen CA, Koppeschaar HP, de Vries WR, Snel YE, Doerga ME, Zelissen PM, Thijssen JH, Blankenstein MA: Visceral adipose tissue is associated with circulating high affinity growth hormone-binding protein. J Clin Endocrinol Metab 1997;82:760–764.
69 Nam SY, Kim KR, Song YD, Lim SK, Lee HC, Huh KB: GH-binding protein in obese men with varying glucose tolerance: Relationship to body fat distribution, insulin secretion and the GH-IGF-I axis. Eur J Endocrinol 1999;140: 159–163.
70 Postel-Vinay MC, Leger J, Sotiropoulos A, Delehaye-Zervas MC, Finidori J, Kelly PA: Regulation of growth hormone-binding protein: Clinical implications. J Pediatr Endocrinol 1993;6:241–244.
71 Baumann G, Shaw MA, Amburn K: Circulating growth hormone binding proteins. J Endocrinol Invest 1994;17:67–81.
72 Peter MA, Winterhalter KH, Boni-Schnetzler M, Froesch ER, Zapf J: Regulation of insulin-like growth factor-I (IGF-I) and IGF-binding proteins by growth hormone in rat white adipose tissue. Endocrinology 1993;133:2624–2631.
73 Vikman K, Isgaard J, Eden S: Growth hormone regulation of insulin-like growth factor-I mRNA in rat adipose tissue and isolated rat adipocytes. J Endocrinol 1991;131:139–145.
74 Gaskins HR, Kim JW, Wright JT, Rund LA, Hausman GJ: Regulation of insulin-like growth factor-I ribonucleic acid expression, polypeptide secretion, and binding protein activity by growth hormone in porcine preadipocyte cultures. Endocrinology 1990;126:622–630.
75 D'Ercole AJ, Stiles AD, Underwood LE: Tissue concentrations of somatomedin C: Further evidence for multiple sites of synthesis and paracrine or autocrine mechanisms of action. Proc Natl Acad Sci USA 1984;81:935–939.
76 Dastot F, Sobrier ML, Duquesnoy P, Duriez B, Goossens M, Amselem S: Alternatively spliced forms in the cytoplasmic domain of the human growth hormone (GH) receptor regulate its ability to generate a soluble GH-binding protein. Proc Natl Acad Sci USA 1996;93:10723–10728.
77 Ross RJ, Esposito N, Shen XY, Von Laue S, Chew SL, Dobson PR, Postel-Vinay MC, Finidori J: A short isoform of the human growth hormone receptor functions as a dominant negative inhibitor of the full-length receptor and generates large amounts of binding protein. Mol Endocrinol 1997;11:265–273.
78 Amit T, Bergman T, Dastot F, Youdim MB, Amselem S, Hochberg Z: A membrane-fixed, truncated isoform of the human growth hormone receptor. J Clin Endocrinol Metab 1997; 82:3813–3817.
79 Shen XY, Holt RI, Miell JP, Justice S, Portmann B, Postel-Vinay MC, Ross RJ: Cirrhotic liver expresses low levels of the full-length and truncated growth hormone receptors. J Clin Endocrinol Metab 1998;83:2532–2538.
80 Wickelgren RB, Landin KL, Ohlsson C, Carlsson LM: Expression of exon 3-retaining and exon 3-excluding isoforms of the human growth hormone-receptor is regulated in an interindividual, rather than a tissue-specific, manner. J Clin Endocrinol Metabol 1995;80: 2154–2157.
81 LaFranchi S, Hanna CE, Torresani T, Schoenle E, Illig R: Comparison of growth hormone binding and metabolic response in rat adipocytes of epididymal, subcutaneous, and retroperitoneal origin. Acta Endocrinol (Copenh) 1985;110:50–55.
82 Bengtsson BA, Eden S, Lonn L, Kvist H, Stokland A, Lindstedt G, Bosaeus I, Tolli J, Sjöström L, Isaksson OG: Treatment of adults with growth hormone (GH) deficiency with recombinant human GH. J Clin Endocrinol Metab 1993;76:309–317.

83 Johannsson G, Marin P, Lonn L, Ottosson M, Stenlof K, Björntorp P, Sjöström L, Bengtsson BA: Growth hormone treatment of abdominally obese men reduces abdominal fat mass, improves glucose and lipoprotein metabolism, and reduces diastolic blood pressure J Clin Endocrinol Metab 1997;82:727–734.
84 Kim KR, Nam SY, Song YD, Lim SK, Lee HC, Huh KB: Low-dose growth hormone treatment with diet restriction accelerates body fat loss, exerts anabolic effect and improves growth hormone secretory dysfunction in obese adults. Horm Res 1999;51:78–84.
85 Frick GP, Leonard JL, Goodman HM: Effect of hypophysectomy on growth hormone receptor gene expression in rat tissues. Endocrinology 1990;126:3076–3082.
86 Kamel A, Margery V, Norstedt G, Thoren M, Lindgren AC, Bronnegard M, Marcus C: Growth hormone (GH) treatment up-regulates GH receptor mRNA levels in adipocytes from patients with GH deficiency and Prader-Willi syndrome. Pediatr Res 1995;38:418–421.
87 Hochberg Z, Barkey RJ, Even L, Peleg I, Youdim MB, Amit T: The effect of human growth hormone therapy on GH binding protein in GH-deficient children. Acta Endocrinol (Copenh) 1991;125:23–27.
88 Fontoura M, Mugnier E, Brauner R, Rappaport R, Postel-Vinay MC: Effect of growth hormone on the low level of growth hormone binding protein in idiopathic short stature. Clin Endocrinol (Oxf) 1992;37:249–253.
89 Morikawa M, Nixon T, Green H: Growth hormone and the adipose conversion of 3T3 cells. Cell 1982;29:783–789.
90 Nixon T, Green H: Contribution of growth hormone to the adipogenic activity of serum. Endocrinology 1984;114:527–532.
91 Hauner H, Löffler G: Adipogenic factors in human serum promote the adipose conversion of 3T3-L1 fibroblasts. Int J Obes 1986;10:323–330.
92 Doglio A, Dani C, Grimaldi P, Ailhaud G: Growth hormone regulation of the expression of differentiation-dependent genes in preadipocyte Ob1771 cells. Biochem J 1986;238:123–129.
93 Deslex S, Negrel R, Ailhaud G: Development of a chemically defined serum-free medium for differentiation of rat adipose precursor cells. Exp Cell Res 1987;168:15–30.
94 Vassaux G, Negrel R, Ailhaud G, Gaillard D: Proliferation and differentiation of rat adipose precursor cells in chemically defined medium: Differential action of anti-adipogenic agents. J Cell Physiol 1994;161:249–256.
95 Tontonoz P, Kim JB, Graves RA, Spiegelman BM: ADD1: A novel helix-loop-helix transcription factor associated with adipocyte determination and differentiation. Mol Cell Biol 1993; 13:4753–4759.
96 Kim JB, Spiegelman BM: ADD1/SREBP1 promotes adipocyte differentiation and gene expression linked to fatty acid metabolism. Genes Dev 1996;10:1096–1107.
97 Elbrecht A, Chen Y, Cullinan CA, Hayes N, Leibowitz MD, Moller DE, Berger J: Molecular cloning, expression and characterization of human peroxisome proliferator activated receptors gamma 1 and gamma 2. Biochem Biophys Res Commun 1996;224:431–437.
98 Tontonoz P, Hu E, Graves RA, Budavari AI, Spiegelman BM: mPPAR gamma 2: Tissue-specific regulator of an adipocyte enhancer. Genes Dev 1994;8:1224–1234.
99 Zhu Y, Qi C, Korenberg JR, Chen XN, Noya D, Rao MS, Reddy JK: Structural organization of mouse peroxisome proliferator-activated receptor gamma (mPPAR gamma) gene: Alternative promoter use and different splicing yield two mPPAR gamma isoforms. Proc Natl Acad Sci USA 1995;92:7921–7925.
100 Brun RP, Kim JB, Hu E, Altiok S, Spiegelman BM: Adipocyte differentiation: A transcriptional regulatory cascade. Curr Opin Cell Biol 1996;8:826–832.
101 Distel RJ, Robinson GS, Spiegelman BM: Fatty acid regulation of gene expression: Transcriptional and post-transcriptional mechanisms. J Biol Chem 1992;267:5937–5941.
102 Moustaid N, Sakamoto K, Clarke S, Beyer RS, Sul HS: Regulation of fatty acid synthase gene transcription. Sequences that confer a positive insulin effect and differentiation-dependent expression in 3T3-L1 preadipocytes are present in the 332 bp promoter. Biochem J 1993;292:767–772.
103 Moustaid N, Sul HS: Regulation of expression of the fatty acid synthase gene in 3T3-L1 cells by differentiation and triiodothyronine. J Biol Chem 1991;266:18550–18554.
104 Hansen LH, Madsen B, Teisner B, Nielsen JH, Billestrup N: Characterization of the inhibitory effect of growth hormone on primary preadipocyte differentiation. Mol Endocrinol 1998;12:1140–1149.
105 Hu E, Kim JB, Sarraf P, Spiegelman BM: Inhibition of adipogenesis through MAP kinase-mediated phosphorylation of PPAR-gamma. Science 1996;274:2100–2103.
106 Hauner H, Rohrig K, Petruschke T: Effects of epidermal growth factor (EGF), platelet-derived growth factor (PDGF) and fibroblast growth factor (FGF) on human adipocyte development and function. Eur J Clin Invest 1995;25:90–96.
107 Smith PJ, Wise LS, Berkowitz R, Wan C, Rubin CS: Insulin-like growth factor-I is an essential regulator of the differentiation of 3T3-L1 adipocytes. J Biol Chem 1988;263: 9402–9408.
108 Schmidt W, Poll-Jordan G, Löffler G: Adipose conversion of 3T3-L1 cells in a serum-free culture system depends on epidermal growth factor, insulin-like growth factor I, corticosterone, and cyclic AMP. J Biol Chem 1990;265:15489–15495.
109 Bonnet F, Vanderschueren-Lodeweyckx M, Eeckels R, Malvaux P: Subcutaneous adipose tissue and lipids in blood in growth hormone deficiency before and after treatment with human growth hormone. Pediatr Res 1974;8: 800–805.
110 Davidson MB: Effect of growth hormone on carbohydrate and lipid metabolism. Endocr Rev 1987;8:115–131.
111 Dietz J, Schwartz J: Growth hormone alters lipolysis and hormone-sensitive lipase activity in 3T3-F442A adipocytes. Metabolism 1991;40:800–806.
112 Adamafio NA, Ng FM: Effects of growth hormone on lipogenesis and glucose oxidation in genetically GH-deficient mice. Mol Cell Endocrinol 1984;37:241–244.
113 Landron D, Guerre-Millo M, Postel-Vinay MC, Lavau M: Relationship between increased binding and insulin-like effects of human growth hormone in adipocytes from young fa/fa rats. Endocrinology 1989;124: 2305–2313.
114 Marcus C, Bolme P, Micha-Johansson G, Margery V, Bronnegard M: Growth hormone increases the lipolytic sensitivity for catecholamines in adipocytes from healthy adults. Life Sci 1994;54:1335–1341.
115 Wabitsch M, Hauner H, Heinze E, Teller W: In vitro effects of growth hormone in adipose tissue. Acta Paediatr 1994;406:48–53.
116 Ridderstrale M, Tornqvist H: Effects of tyrosine kinase inhibitors on tyrosine phosphorylations and the insulin-like effects in response to human growth hormone in isolated rat adipocytes. Endocrinology 1996;137:4650–4656.
117 Shimizu M, Torti F, Roth RA: Characterization of the insulin and insulin-like growth factor receptors and responsitivity of a fibroblast/adipocyte cell line before and after differentiation. Biochem Biophy Res Commun 1986;137:552–558.
118 O'Neal DN, Kalfas A, Dunning PL, Christopher MJ, Sawyer SD, Ward GM, Alford FP: The effect of 3 months of recombinant human growth hormone (GH) therapy on insulin and glucose-mediated glucose disposal and insulin secretion in GH-deficient adults: A minimal model analysis. J Clin Endocrinol Metab 1994;79:975–983.
119 Rizza RA, Mandarino LJ, Gerich JE: Effects of growth hormone on insulin action in man. Mechanisms of insulin resistance, impaired suppression of glucose production, and impaired stimulation of glucose utilization. Diabetes 1982;31:663–669.
120 Maloff BL, Levine JH, Lockwood DH: Direct effects of growth hormone on insulin action in rat adipose tissue maintained in vitro. Endocrinology 1980;107:538–544.
121 Schwartz J: Growth hormone directly alters glucose utilization in 3T3 adipocytes. Biochem Biophys Res Commun 1984;125:237–243.
122 Schwartz J, Carter-Su C: Effects of growth hormone on glucose metabolism and glucose transport in 3T3-F442A cells: Dependence on cell differentiation. Endocrinology 1988;122: 2247–2256.

123 Tai PK, Liao JF, Chen EH, Dietz J, Schwartz J, Carter-Su C: Differential regulation of two glucose transporters by chronic growth hormone treatment of cultured 3T3-F442A adipose cells. J Biol Chem 1990;265:21828–21834.
124 Kilgour E, Baldwin SA, Flint DJ: Divergent regulation of rat adipocyte GLUT1 and GLUT4 glucose transporters by GH. J Endocrinol 1995;145:27–33.
125 Smith TR, Elmendorf JS, David TS, Turinsky J: Growth hormone-induced insulin resistance: Role of the insulin receptor, IRS-1, GLUT-1, and GLUT-4. Am J Physiol 1997;272:E1071–E1079.
126 Brook CG: Effect of human growth hormone treatment on adipose tissue in children. Arch Dis Child 1973;48:725–728.
127 Beauville M, Harant I, Crampes F, Rivière D, Tauber MT, Tauber JP, Garrigues M: Effect of long-term rhGH administration in GH-deficient adults on fat cell epinephrine response. Am J Physiol 1992;263:E467–E472.
128 Kamel A, Norgren S, Eliman A, Danielsson P, Marcus C: Effect of growth hormone treatment in obese prepubertal boys. J Clin Endocrinol Metab 2000;85:1412–1419.
129 Yang S, Xu X, Björntorp P, Eden S: Additive effects of growth hormone and testosterone on lipolysis in adipocytes of hypophysectomized rats. J Endocrinol 1995;147:147–152.
130 Yang S, Björntorp P, Liu X, Eden S: Growth hormone treatment of hypophysectomized rats increases catecholamine-induced lipolysis and the number of beta-adrenergic receptors in adipocytes: No differences in the effects of growth hormone on different fat depots. Obes Res 1996;4:471–478.
131 Kamel A, Norgren S, Lindgren AC, Luthman H, Arner P, Marcus C: Effect of growth hormone treatment on insulin action in adipocytes from children with Prader-Willi syndrome. Eur J Endocrinol 1998;138:510–516.
132 Eckel RH, Yost TJ: Weight reduction increases adipose tissue lipoprotein lipase responsiveness in obese women. J Clin Invest 1987;80:992–997.
133 Richelsen B, Pedersen SB, Borglum JD, Möller-Pedersen T, Jörgensen J, Jorgensen JO: Growth hormone treatment of obese women for 5 wk: Effect on body composition and adipose tissue LPL activity. Am J Physiol 1994;266:E211–E216.
134 Ottosson M, Vikman-Adolfsson K, Enerbäck S, Elander A, Björntorp P, Eden S: Growth hormone inhibits lipoprotein lipase activity in human adipose tissue. J Clin Endocrinol Metab 1995;80:936–941.
135 Rosenbaum M, Gertner JM, Leibel RL: Effects of systemic growth hormone (GH) administration on regional adipose tissue distribution and metabolism in GH-deficient children. J Clin Endocrinol Metab 1989;69:1274–1281.
136 Flint DJ, Gardner MJ: Influence of growth hormone deficiency on growth and body composition in rats: Site-specific effects upon adipose tissue development. J Endocrinol 1993;137:203–211.
137 Bengtsson BA: The consequences of growth hormone deficiency in adults. Acta Endocrinol (Copenh) 1993;128(suppl 2):2–5.
138 Beshyah SA, Johnston DG: Cardiovascular disease and risk factors in adults with hypopituitarism. Clin Endocrinol (Oxf) 1999;50:1–15.
139 Rosen T, Johannsson G, Johansson JO, Bengtsson BA: Consequences of growth hormone deficiency in adults and the benefits and risks of recombinant human growth hormone treatment. Horm Res 1995;43:93–99.
140 Bengtsson BA, Johannsson G: Effect of growth hormone therapy on early atherosclerotic changes in GH-deficient adults. Lancet 1999;353:1898–1899.
141 Johannsson G, Bengtsson BA: Growth hormone and the metabolic syndrome. J Endocrinol Invest 1999;22(5 suppl):41–46.
142 Forbes GB: Weight loss during fasting: Implications for the obese. Am J Clin Nutr 1970;23:1212–1219.
143 Snyder DK, Clemmons DR, Underwood LE: Treatment of obese, diet-restricted subjects with growth hormone for 11 weeks: Effects on anabolism, lipolysis, and body composition. J Clin Endocrinol Metab 1988;67:54–61.
144 Jörgensen JO, Pedersen SB, Borglum J, Möller N, Schmitz O, Christiansen JS, Richelsen B: Fuel metabolism, energy expenditure, and thyroid function in growth hormone-treated obese women: A double-blind placebo-controlled study. Metabolism 1994;43:872–877.
145 Tagliaferri M, Scacchi M, Pincelli AI, Berselli ME, Silvestri P, Montesano A, Ortolani S, Dubini A, Cavagnini F: Metabolic effects of biosynthetic growth hormone treatment in severely energy-restricted obese women. Int J Obes Relat Metab Disord 1998;22:836–841.
146 Snyder DK, Underwood LE, Clemmons DR: Persistent lipolytic effect of exogenous growth hormone during caloric restriction. Am J Med 1995;98:129–134.
147 Snyder DK, Underwood LE, Clemmons DR: Anabolic effects of growth hormone in obese diet-restricted subjects are dose dependent. Am J Clin Nutr 1990;52:431–437.
148 Clemmons DR, Snyder DK, Williams R, Underwood LE: Growth hormone administration conserves lean body mass during dietary restriction in obese subjects. J Clin Endocrinol Metab 1987;64:878–883.
149 Cuneo RC, Salomon F, Wilmshurst P, Byrne C, Wiles CM, Hesp R, Sonksen PH: Cardiovascular effects of growth hormone treatment in growth-hormone-deficient adults: Stimulation of the renin-aldosterone system. Clin Sci 1991;81:587–592.
150 Rudling M, Norstedt G, Olivecrona H, Reihner E, Gustafsson JA, Angelin B: Importance of growth hormone for the induction of hepatic low density lipoprotein receptors. Proc Natl Acad Sci USA 1992;89:6983–6987.
151 Chong PK, Jung RT, Scrimgeour CM, Rennie MJ, Paterson CR: Energy expenditure and body composition in growth hormone deficient adults on exogenous growth hormone. Clin Endocrinol (Oxf) 1994;40:103–110.

Hormone Research

Author Index Vol. 53, Suppl. 1, 2000

Subject Index Vol. 53, Suppl. 1, 2000

KARGER
Fax + 41 61 306 12 34
E-Mail karger@karger.ch
www.karger.com

Accessible online at:
www.karger.com/journals/hre